# PRACTICAL BOOKLET 1: Commands Windows network and AD DS

Copyright © 25-09-2016 by Baldomero Sánchez Pérez

All rights reserved. This book or any part there of may not be reproduced or used in any form without the express written permission of the publisher, except for the use of postage quotes in a review of a book or journal.

First Drive: 2016

ISBN:   978-1-326-79823-9

Editorial: LULU.COM

www.baldoweb.net

Baldomero Sánchez Pérez

Technical Teacher Training
Systems and Computer Applications

Informatics Engineer
Pontifical University of Salamanca

This book is dedicated to the students, who based their effort and enthusiasm, they achieve every year started the training cycles degree as Technicians Microcomputer Systems and Networks, such as performing cycles Degree in Management Computer systems and Networks.

And especially these large computer professionals, roommates, who quietly and with its spirit of professional work are current on innovations Informatics, getting students to achieve the necessary knowledge for professional and personal development.

Do not forget the most important thing in my life is my family, which I appreciate your support and enthusiasm that allow to develop best my work.

*"I do not know, it has become not know yet"*

*Bill Gates*

*PRACTICAL BOOKLET 1: Commands Windows network and AD DS*

*CONTENT*

PREFACE ......................................................................................................................................... 8
PRACTICE 1: Perform Network Configuration. ............................................................................ 10
PRACTICE 2: Configure Network structure: UO, groups and users. ........................................... 28
PRACTICE 3: Create a user, O, groups, users within groups. DSADD ......................................... 43
PRACTICE 4: Permissions Windows Users. ................................................................................. 44
PRACTICE 5: Display AD DS domain information ....................................................................... 46
PRACTICE 6: Display object information and network drives domain. ..................................... 52
PRACTICE 7: Display and set the partitions and quotas, in AD DS system. ............................... 58
PRACTICE 8: Restrictions Windows key level in AD DS. ............................................................. 62
PRACTICE 9: Analyze and see the protocols and processes running sessions .......................... 68
PRACTICE 10: Test the DNS server ............................................................................................. 85
PRACTICE 11: Routing Tables. ROUTER ..................................................................................... 90
ATTACHMENTS: RED commands and AD DS .............................................................................. 93
GLOSARIO ................................................................................................................................. 118
REFERENCE COMMANDS .......................................................................................................... 120

# PREFACE

This book is called practical booklet, because in it basic practices of Windows commands networks that handle both the client side and the server side (Windows Server) are collected, collects some of the concepts of teaching programs, which they are incorporated in different work units of different modules of training cycles (average grade as the various higher degrees in computer science). There is no classification by work units or by chapter, as it is a sequential compendium of practices developed both in Active Directory (Windows Server 2012 R2) in a Windows 10 Professional client.

It's booklet and no notebook, your organization is basic practices to handle the following commands: hostname, getmac, ping, pathping, tracert, nbstat, arp, route, ipconfig, nslookup, netstat, whoami, dsadd, icacls, dsacls, dsget , dsquery, ...

Goal are practical command that must be accompanied by a relevant to the con-cepts of networks to be tested and analyzed in the network in case of ignorance, since it is based on the premise explanation that the user has the basic knowledge of networking and communication theory.

It is a totally valid for any IT professional book, want to have a guide network command execution and analysis of the results obtained, as a practical reference.

The Organization of the book is based on 11 practices that can be implemented and used independently or in combination or in parts.

The practices are organized, first the objective to be achieved arises, corresponds to the title of the practice then is the description of basic knowledge for the development of practice, performance requirements, management and steps to be continue to develop, they are arranged numerically. At each step there Alphabetic points, with the expla-nation of each option or modifier and its results, or the combination of several options. The practices contain illustra-tions or text results obtained in the PROMPT, the results come from servers running on virtual machines, VirtualBox and Hyper-V for Windows 10.

Supplements contain explanatory practices of its accomplishment or prior knowledge relating to its development at a practical level. They are accompanied by bullets or clarification of its development, reflected-das in different colors: Blue explanations of knowledge for its development, Orange explanatory notes or important requirements or precautions, Green clarification of parameters and values to be taken into account in implementation.

The methodology used in the development of practices is a constructive methodology. The complement to an ini-tial learning behaviorist intended. Although it can be used for distance learning or self-study or professional consulta-tion network technicians.

For its development has been used genuine Microsoft software © 2012 R2 Windows Server, Windows 7 Pro-fessional and Windows 10 Professional, MSDN under license. The goal is not to modify or plagiarize, only disclose the contents as didactic learning, under the trademarks, under current legislation. The illustrations images or graphics used comes from some Web with record ©, as such are referenced location without any modification thereof, such as CISCO, ORACLE or Wikipedia in turn are referenced the Web, where it has been extracted command information such as eg Microsoft .: http://ss64.com/nt/, https: //technet.miscrosoft.com,... They are located at the end, in the web References point.

# PRACTICE 1: Perform Network Configuration.

**DESCRIPTION:**

a) Introduction.

Its importance is important to have knowledge of administration, since today we are in the environment that everything is communication because commands allow us to communicate with other computers (HOST).

Network commands, its usefulness is to detect the proper operation and possible failures or problems in a local area network and any other area (including Internet).

His execution is done from the console prompt (CMD), PowerShell or from any application that performs invocation system commands or objects, such as from Visual Basic Scripting Host, NetLogon, MSIC or any other application.

The deep network commands, knowledge means their only use in combination with the Po-werShell commands, administrator-level facilities in the CORE Windows 2008, 2012, 2016 Server.

b) Analyze Network configuration.

A network has as main element of a physical device communication, mainly the network card or other devices (Bluetooth ... Modem "obsolete" ...

Every physical network device, or any network interface card (NIC: Network Interface Control). It has direct communication with the computer hardware is a direct interaction with the operating system, downloading the microprocessor of its management, is reflected in the area of computer RAM (DMA: Direct Memory Access) and consisting of: the data addresses and recording access control device, together with the (s) interrupt (s) handling. This allows an interactive operation between the hardware system and NIC.

All computer on a network is identified by its name team for their MAC, both must be unique.

If you also use the TCP / IP communications on a network it implies that there can not be two teams with the same IP or the same name.

Windows installation, assign the default workgroup WORKGROUP or WORKGROUP and does not establish IP direc-tion, being active by default using both IPv4 and IPv6 protocols, in order to allow an IP address is assigned by default a DHCP server. Assigns a random default name.

## STEP 1: Check the computer name. HOSTNAME

Display the name of the local computer. In the process of installing Windows, by default is assigned a name to the computer. If the user has not changed is the name that appears with the following command. To change the name in Windows 2012 Server (Start-> Settings-> System-> About ... -> Rename), Windows 10, My Computer (right button) -> Properties-> Change Settings).

a) Display Team name from CMD.

**HOSTNAME**
```
C:\Users\aprendiz>hostname
i7-PC
```
b) Rename the computer from CMD.
Exactly it works on both Windows 2012 Server or Windows 10..
b.1) From the PROMPT. Writes **sysdm.cpl** + [ENTER]
C:\Users\aprendiz>sysdm.cpl
b.2) From the graphical environment, to run.
Key in Windows + X
Execute: sysdm.cpl + [ACEPTAR]

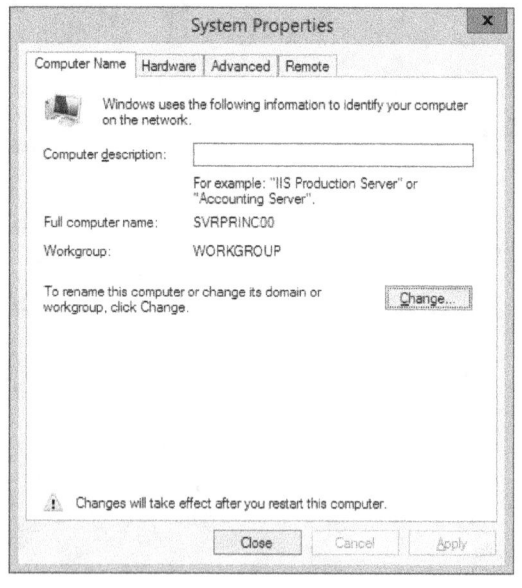

## STEP 2: Check the MAC address of the network cards. GETMAC

**What is MAC?**

The MAC (Media Access Control; "media access control") address is a 48-bit identifier (6 blo-ques hexadecimal) which corresponds uniquely to a card or network device. It is also known as physical address, and is unique for each network device, there can not be two equal-level manufacturer.

It consists of two parts, each consisting of 24 bits.

OUI: It is determined and set by the IEEE (the first 24 bits) and the manufacturer (the last 24 bits) using the Organizationally Unique Identifier.

NIC: The 24 bits that identify the network device, typical of a manufacturer (OUI), and is unique in its manufacture. There can be no two are alike.

$2^{24}$ bits= 16.777.216
$2^{48}$ bits= 281.474.976.710.656

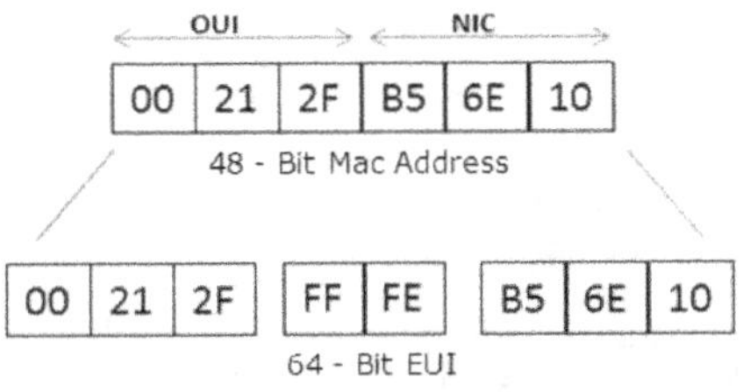

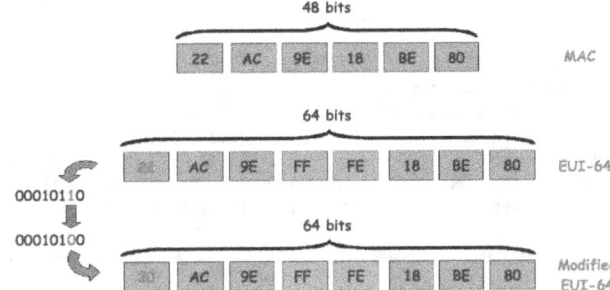

Ilustración 1: MAC Address comes from Wikipedia

The EUI-64 IPv6 address format is obtained through the MAC address of 48 bits. The MAC address is separated first two 24 bits, one of which OUI (Organizationally Unique Identifier) and the other is the specific NIC. The 16-bit 0xFFFE is then inserted between these two 24-bit EUI 64 address bits. IEEE has chosen FFFE as a reserved value that can only appear in EUI-64 generated from one EUI-48 MAC address.

The Most Significant Byte, the seventh bit of the left, or the universal / local bit (U / L), has to be reversed. This bit identifies whether this interface identifier is universally or locally administered:

- 0: locally administered address.
- 1: The address is unique worldwide.

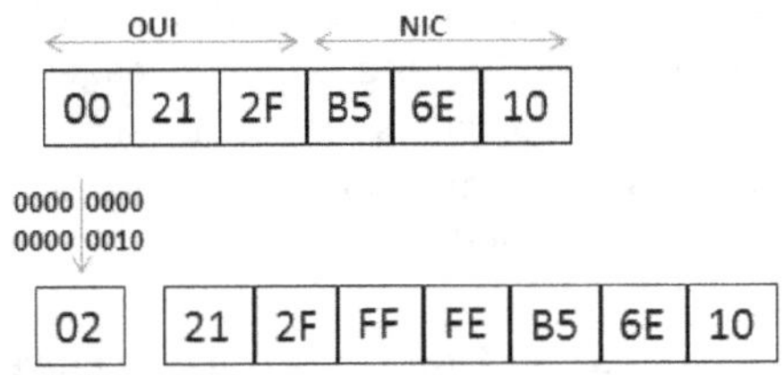

Should be highlighted in the OUI part, the globally unique addresses assigned by the IEEE has always been set to 0, while the addresses created locally configured 1. Therefore, when the bit is reversed, maintains its original scope ( global unique address remains only global and vice versa). The reason for investing can be found in section 2.5.1 RFC4291.

Ilustración 2: http://www.openwall.com

Figure 3 Explanation of local/universal positición 2 Bit, this image comes from CISCO ranks second bit to the right.
Image comes from
https://supportforums.cisco.com/document

## STEP 2.1. Consult local cards.
a) Default query.

```
C:\Windows\system32>GETMAC

Physical Address    Transport Name
=================   =======================================================
68-5D-43-E2-34-ED   \Device\Tcpip_ 6C3F6FE1-C095-434B-9A2A-9B1DB3D61790}
5C-F9-DD-40-96-17   Media disconnected
N/A                 Hardware absent
68-5D-43-E2-34-F1   Media disconnected
5C-F9-DD-40-96-17   Media disconnectedDetailed Consultation cards

C:\Windows\system32>GETMAC  /V

Connection Name  Network Adapter   Physical Address    Transport Name
===============  ===============   =================   ==============================================
Conexión de red  Intel(R) Centri   68-5D-43-E2-34-ED   \Device\Tcpip_{6C3F6FE1-C095-434B-9A2A-9B1DB3D61790}
Ethernet         Realtek PCIe FE   5C-F9-DD-40-96-17   Medios desconectados
VirtualBox Host  VirtualBox Host   N/A                 Hardware ausente
Conexión de red  Bluetooth Devic   68-5D-43-E2-34-F1   Medios desconectados
vEthernet (Mi t  Hyper-V Virtual   5C-F9-DD-40-96-17   Medios desconectados
```

It is located at Layer 2 of the OSI model using one of the three numerations handled by IEEE: MAC-48, EUI-48 and EUI-64.

b) Check the output format list, table and CSV.

```
C:\Windows\system32>GETMAC  /V /FO LIST

Connection Name  Network Adapter   Physical Address    Transport Name
===============  ===============   =================   ==============================================
Conexión de red  Intel(R) Centri   68-5D-43-E2-34-ED   \Device\Tcpip_{6C3F6FE1-C095-434B-9A2A-9B1DB3D61790}
Ethernet         Realtek PCIe FE   5C-F9-DD-40-96-17   Medios desconectados
VirtualBox Host  VirtualBox Host   N/A                 Hardware ausente
Conexión de red  Bluetooth Devic   68-5D-43-E2-34-F1   Medios desconectados
vEthernet (Mi t  Hyper-V Virtual   5C-F9-DD-40-96-17   Medios desconectados

C:\Windows\system32>GETMAC  /V /FO TABLE

Connection Name  Network Adapter   Physical Address    Transport Name
===============  ===============   =================   ==============================================
Conexión de red  Intel(R) Centri   68-5D-43-E2-34-ED   \Device\Tcpip_{6C3F6FE1-C095-434B-9A2A-9B1DB3D61790}
```

```
Ethernet              Realtek PCIe FE  5C-F9-DD-40-96-17  Medios desconectados
VirtualBox Host VirtualBox Host N/A                        Hardware ausente
Conexión de red Bluetooth Devic 68-5D-43-E2-34-F1          Medios desconectados
vEthernet (Mi t Hyper-V Virtual 5C-F9-DD-40-96-17          Medios desconectados

C:\Windows\system32>GETMAC   /v /fo csv
"Connection Name","Network Adapter","Physical Address","Transport Name"
"Conexión de red inalámbrica","Intel(R) Centrino(R) Wireless-N 2230","68-5D-43-E2-34-
ED","\Device\Tcpip_{6C3F6FE1-C095-434B-9A2A-9B1DB3D61790}"
"Ethernet","Realtek PCIe FE Family Controller","5C-F9-DD-40-96-17","Medios desconectados"
"VirtualBox Host-Only Network","VirtualBox Host-Only Ethernet Adapter","N/A","Hardware ausente"
"Conexión de red Bluetooth 2","Bluetooth Device (Personal Area Network)","68-5D-43-E2-34-F1","Medios desco-
nectados"
"vEthernet (Mi tarjeta de red i7)","Hyper-V Virtual Ethernet Adapter","5C-F9-DD-40-96-17","Medios desconecta-
dos"
```

## STEP 3: Check network cards a server from a workstation. Getmac

The steps to follow in order to be able to check or analyze a MAC, from one job to another network equipment, in the example is performed on a Windows 2012 Server are established. The getmac command must have access to the RPC procotolo, this means that you must enable the RPC service on the firewall.

First the state "Remote Server Manager functionality" is checked, from Powershell.

Second firewall access the graphical environment, and entry rules are set up, enabling remote management of RPC services.

Third from a client, Windows 10, access the server so authenticated with username and password and com-MAC test server from the client. [Point e)]

a) Commands WEB network from the PowerShell executed only on a Windows 2012 Server.
   By command from a command prompt:
```
C:\Windows\system32> POWERSHELL
Windows PowerShell
Copyright (C) 2013 Microsoft Corporation. Todos los derechos reservados.

PS C:\Windows\system32> Configure-SMRemoting -get
Remote Server Manager functionality is enabled

PS C:\Windows\system32> Configure-SMRemoting -disable
Remote Server Manager functionality is now disabled:
  Remote access disabled.

PS C:\Windows\system32> Configure-SMRemoting -enable
Remote Server Manager functionality is now enabled: Enable remote access.
```
> Configure-SMRemoting -get, It shows the state
> Configure-SMRemoting -enable
> Configure-SMRemoting - disable

b) Allows access to the server for this we must enable the firewall.
   Server is enabled on the FIREWALL

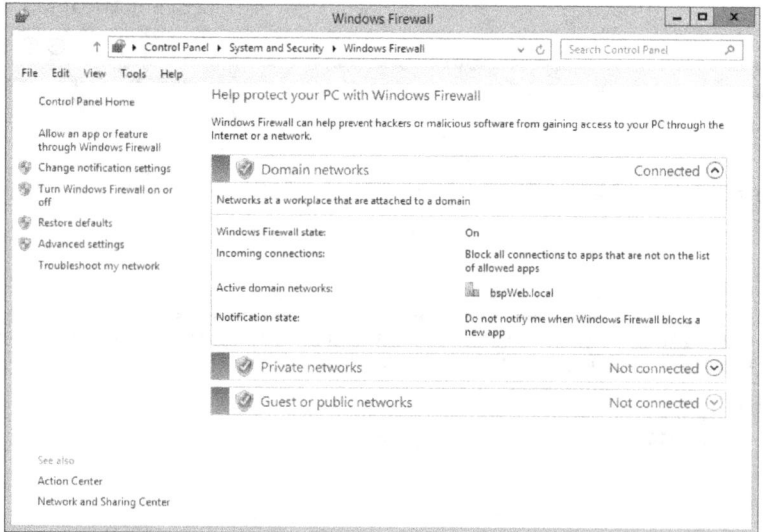

c) In Advanced settings in the menu on the left, and we agree to the rules.
   Inbound Rules

d) Within the rules we are looking for we enable / disable Rule (Right click on the field el-gable desple appears). You must activate remote administration on Windows services. RPC.

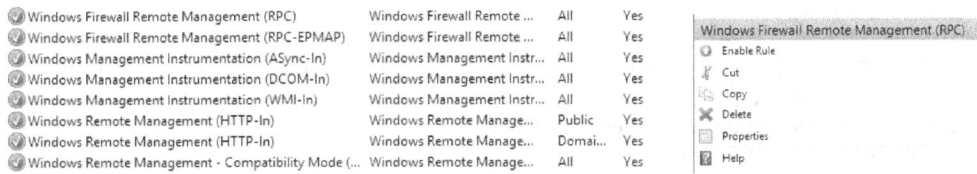

d.1)  Activated by double-clicking on the firewall permission.

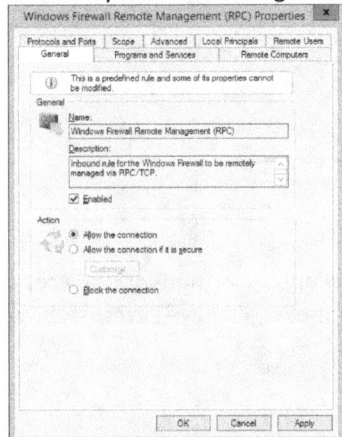

e)  Check from the Windows client if you have access to the server, after remote identification (RPC).
```
Z:\>GETMAC /s SVRPRINC00 /NH /V
Type the password for SAUCES\baldo:*********

Error: login: unknown user name or incorrect password.
```

f)  Access to the server name plus the name of the user who want to connect. If both are correct we request the password, and shows the information on the card (s) requested network (s).
```
Z:\>GETMAC /s SVRPRINC00 /U Administrador
Type the password for Administrator:*********


Physical address    Transport Name
=================== ============================================================
08-00-27-F4-EA-76   \Device\Tcpip_{25AABFF4-914D-41F7-8E22-BBE28F9ADAFE}
```

g)  Access the server with a username and password, shows the network card.
```
Z:\>GETMAC /s SVRPRINC00 /U Administrador  /P Aaa111!!!

Physical address    Transport Name
=================== ============================================================
08-00-27-F4-EA-76   \Device\Tcpip_{25AABFF4-914D-41F7-8E22-BBE28F9ADAFE}
```

h)  Access the server with a username and password, shows the network card, without displaying the table header format.
```
Z:\>GETMAC /s SVRPRINC00 /U Administrador  /P Aaa111!!! /NH

08-00-27-F4-EA-76   \Device\Tcpip_{25AABFF4-914D-41F7-8E22-BBE28F9ADAFE}
```

i)  Access the server with a username and password, shows the network card, formatted list.
```
Z:\>GETMAC /s SVRPRINC00 /U Administrador  /P Aaa111!!! /FO list

Physical address    Transport Name
=================== ============================================================
08-00-27-F4-EA-76   \Device\Tcpip_{25AABFF4-914D-41F7-8E22-BBE28F9ADAFE}
```

j)  Access the server with a username and password, shows the network card, in list format, with complete information.
```
Z:\>GETMAC /s SVRPRINC00 /U Administrador  /P Aaa111!!! /FO list  /v

Connection Name  Network Adapter   Physical Address    Transport Name
===============  ================  ==================  ========================================
Ethernet         Adaptador de es   08-00-27-F4-EA-76   \Device\Tcpip_{25AABFF4-914D
-41F7-8E22-BBE28F9ADAFE}
```

k)  Access the server with a username and password, shows the network card, with table format, with complete information.
```
Z:\>GETMAC /s SVRPRINC00 /U Administrador  /P Aaa111!!! /FO table

Physical address    Transport Name
=================== ============================================================
08-00-27-F4-EA-76   \Device\Tcpip_{25AABFF4-914D-41F7-8E22-BBE28F9ADAFE}
```

l)  Access the server with a username and password, shows the network card, with table format, showing the header information of the table.
```
Z:\>GETMAC /s SVRPRINC00 /U Administrador  /P Aaa111!!! /FO table /NH

08-00-27-F4-EA-76   \Device\Tcpip_{25AABFF4-914D-41F7-8E22-BBE28F9ADAFE}
```

m)  Access the server with a username and password, shows the network card, with table format, with complete information.
```
Z:\>GETMAC /s SVRPRINC00 /U Administrador  /P Aaa111!!! /FO CSV
```

```
" Physical address","Transport Name"
"08-00-27-F4-EA-76","\Device\Tcpip_{25AABFF4-914D-41F7-8E22-BBE28F9ADAFE}"
```

n) Access the server with a user name and password without specifying the request and the cancellation of command execution occurs (the result requested in list format).
```
Z:\>GETMAC /s SVRPRINC00 /U bspWeb\administrador    /FO LIST
Type the password for bspWeb\administrador:^C
```

o) Access the server with a user name and password without specifying the request and prior to execution, shows the network card, with LIST format.
```
Z:\>GETMAC /s SVRPRINC00 /U bspWeb.local\administrador    /FO LIST
Type the password for bspWeb.local\administrador:*********

Physical address        Transport Name
===================     ==========================================================
08-00-27-F4-EA-76       \Device\Tcpip_{25AABFF4-914D-41F7-8E22-BBE28F9ADAFE}
```

## STEP 4: Diagnose network card or media. PING

It is used to check for physical connection, and also communicate with another computer or network device. Is en-route a number of data packets to the address indicated and at the end we will check if these messages have successfully reached their destination (time and number). For proper operation should be allowed to use firewalls.

### STEP 4.1: From a Windows 7 Professional client to a Windows Server 2012 R2

a) Get help on the command line.
```
Z:\>PING /?
```

b) Normal Ping an IP address.
```
Z:\>PING    192.168.5.240

Pinging 192.168.5.240 with 32 bytes of data:
Reply from 192.168.5.240: bytes = 32 time <1ms TTL = 128
Reply from 192.168.5.240: bytes = 32 time <1ms TTL = 128
Reply from 192.168.5.240: bytes = 32 time <1ms TTL = 128
Reply from 192.168.5.240: bytes = 32 time <1ms TTL = 128

Ping statistics for 192.168.5.240:
    Packets: Sent = 4, Received = 4, Lost = 0
    (0% loss),
Approximate round trip times in milliseconds:
    Minimum = 0ms, Maximum = 0ms, Average = 0ms
```

c) Check operation of the network card. The Ping the card itself; localhost (127.0.0.1), is as if the Tx output to the input of Rx card will bypass itself.
```
C:\Windows\system32>PING localhost
Pinging IS31W7PR.sauces.local [:: 1] with 32 bytes of data:
Reply from :: 1: time <1m
Reply from :: 1: time <1m
Reply from :: 1: time <1m
Reply from :: 1: time <1m

Ping statistics for :: 1:
    Packets: Sent = 4, Received = 4, Lost = 0
    (0% loss),
Approximate round trip times in milliseconds:
    Minimum = 0ms, Maximum = 0ms, Average = 0ms

C:\Windows\system32>PING 127.0.0.1

Pinging 127.0.0.1 with 32 bytes of data:
Reply from 127.0.0.1: bytes = 32 time <1ms TTL = 128
Reply from 127.0.0.1: bytes = 32 time <1ms TTL = 128
Reply from 127.0.0.1: bytes = 32 time <1ms TTL = 128
Reply from 127.0.0.1: bytes = 32 time <1ms TTL = 128

Ping statistics for 127.0.0.1:
    Packets: Sent = 4, Received = 4, Lost = 0
    (0% loss),
Approximate round trip times in milliseconds:
    Minimum = 0ms, Maximum = 0ms, Average = 0ms
```

d) Ping a specific host.
```
Z:\>PING -t 192.168.5.240

Pinging 192.168.5.240 with 32 bytes of data:
Reply from 192.168.5.240: bytes = 32 time <1ms TTL = 128
Reply from 192.168.5.240: bytes = 32 time <1ms TTL = 128
Reply from 192.168.5.240: bytes = 32 time <1ms TTL = 128
Reply from 192.168.5.240: bytes = 32 time <1ms TTL = 128
Reply from 192.168.5.240: bytes = 32 time <1ms TTL = 128
```

> **PING:** The origin is the sound of submarines. Another hobby is Packet Internet Groper, "Finder or rastreador packet networks"
> 
> This tool enables the administrator to display the MAC address for network adapters on a system.
> 
> When you run the Ping command request, the local host sends an ICMP message, embedded in an IP packet. The ICMP request message includes, besides the type of message and the code of an ID number and a sequence of numbers, 32 bits, which must match the ICMP response message; plus an optional data space. As ICMP protocol is not based on a transport layer protocol such as TCP or UDP and uses no application layer protocol. (Ref. Wikipedia)

```
Reply from 192.168.5.240: bytes = 32 time <1ms TTL = 128
Reply from 192.168.5.240: bytes = 32 time <1ms TTL = 128
Reply from 192.168.5.240: bytes = 32 time <1ms TTL = 128
Reply from 192.168.5.240: bytes = 32 time <1ms TTL = 128
Reply from 192.168.5.240: bytes = 32 time <1ms TTL = 128
Reply from 192.168.5.240: bytes = 32 time <1ms TTL = 128

Ping statistics for 192.168.5.240:
    Packets: Sent = 11, received = 11, Lost = 0
    (0% loss),
Approximate round trip times in milliseconds:
    Minimum = 0ms, Maximum = 0ms, Average = 0ms
Control-C
^C
```

e) Resolve addresses of a host. From a domain server.
```
C:\Windows\system32>PING -a www.google.es

Pinging www.google.es [216.58.211.227] with 32 bytes of data:
Reply from 216.58.211.227: bytes = 32 time = 15ms TTL = 54
Reply from 216.58.211.227: bytes = 32 time = 17ms TTL = 54
Reply from 216.58.211.227: bytes = 32 time = 18ms TTL = 54
Reply from 216.58.211.227: bytes = 32 time = 14ms TTL = 54

Ping statistics for 216.58.211.227:
    Packets: Sent = 4, Received = 4, Lost = 0
    (0% loss),
Approximate round trip times in milliseconds:
    Minimum = 14ms, Maximum = 18ms, average =   16ms
```

f) Specify an address and independently analyze the -a option.
```
Z:\>PING  -a 192.168.5.240

Pinging SVRPRINC00 [192.168.5.240] with 32 bytes of data:
Reply from 192.168.5.240: bytes = 32 time <1ms TTL = 128
Reply from 192.168.5.240: bytes = 32 time <1ms TTL = 128
Reply from 192.168.5.240: bytes = 32 time <1ms TTL = 128
Reply from 192.168.5.240: bytes = 32 time <1ms TTL = 128

Ping statistics for 192.168.5.240:
    Packets: Sent = 4, Received = 4, Lost = 0
    (0% loss),
Approximate round trip times in milliseconds:
    Minimum = 0ms, Maximum = 0ms, Average = 0ms
```

g) Ping a specific host. Independently until it stops (-t).
```
Z:\>PING -t -a 192.168.5.240

Pinging SVRPRINC00 [192.168.5.240] with 32 bytes of data:
Reply from 192.168.5.240: bytes = 32 time <1ms TTL = 128
Reply from 192.168.5.240: bytes = 32 time <1ms TTL = 128
Reply from 192.168.5.240: bytes = 32 time <1ms TTL = 128
Reply from 192.168.5.240: bytes = 32 time <1ms TTL = 128
Reply from 192.168.5.240: bytes = 32 time <1ms TTL = 128
Reply from 192.168.5.240: bytes = 32 time <1ms TTL = 128
Reply from 192.168.5.240: bytes = 32 time <1ms TTL = 128
Reply from 192.168.5.240: bytes = 32 time <1ms TTL = 128
Reply from 192.168.5.240: bytes = 32 time <1ms TTL = 128
Reply from 192.168.5.240: bytes = 32 time <1ms TTL = 128

Ping statistics for 192.168.5.240:
    Packets: Sent = 10, received = 10, Lost = 0
    (0% loss),
Approximate round trip times in milliseconds:
    Minimum = 0ms, Maximum = 0ms, Average = 0ms
Control-C
^C
```

h) Check the resolution of a host for a number of echo requests (6 echo requests).
```
Z:\>PING   -a -n 6  192.168.5.240

Pinging SVRPRINC00 [192.168.5.240] with 32 bytes of data:
Reply from 192.168.5.240: bytes = 32 time <1ms TTL = 128
Reply from 192.168.5.240: bytes = 32 time <1ms TTL = 128
Reply from 192.168.5.240: bytes = 32 time <1ms TTL = 128
Reply from 192.168.5.240: bytes = 32 time <1ms TTL = 128
Reply from 192.168.5.240: bytes = 32 time <1ms TTL = 128
Reply from 192.168.5.240: bytes = 32 time <1ms TTL = 128

Ping statistics for 192.168.5.240:
    Packets: Sent = 6, received = 6, Lost = 0
```

```
            (0% loss),
    Approximate round trip times in milliseconds:
        Minimum = 0ms, Maximum = 0ms, Average = 0ms

    Z:\>PING  -a -n 6  8.8.8.8

    Pinging google-public-dns-a.google.com [8.8.8.8] with 32 bytes of data:
    Reply from 8.8.8.8: bytes = 32 time = 16ms TTL = 51
    Reply from 8.8.8.8: bytes = 32 time = 16ms TTL = 51
    Reply from 8.8.8.8: bytes = 32 time = 16ms TTL = 51
    Reply from 8.8.8.8: bytes = 32 time = 16ms TTL = 51
    Reply from 8.8.8.8: bytes = 32 time = 16ms TTL = 51
    Reply from 8.8.8.8: bytes = 32 time = 16ms TTL = 51

    Ping statistics for 8.8.8.8:
        Packets: Sent = 6, received = 6, Lost = 0
        (0% loss),
    Approximate round trip times in milliseconds:
        Minimum = 16ms, Maximum = 16ms, average = 16ms
```

i) Cancel the execution of ping ^ + C, which has fixed a number of request 30.
```
    Z:\>PING  -n 30  8.8.8.8

    Pinging 8.8.8.8 with 32 bytes of data:
    Reply from 8.8.8.8: bytes = 32 time = 16ms TTL = 51
    Reply from 8.8.8.8: bytes = 32 time = 16ms TTL = 51
    Reply from 8.8.8.8: bytes = 32 time = 16ms TTL = 51
    Reply from 8.8.8.8: bytes = 32 time = 16ms TTL = 51
    Reply from 8.8.8.8: bytes = 32 time = 16ms TTL = 51
    Reply from 8.8.8.8: bytes = 32 time = 16ms TTL = 51
    Reply from 8.8.8.8: bytes = 32 time = 16ms TTL = 51
    Reply from 8.8.8.8: bytes = 32 time = 16ms TTL = 51
    Reply from 8.8.8.8: bytes = 32 time = 16ms TTL = 51

    Ping statistics for 8.8.8.8:
        Packets: Sent = 9, received = 9, Lost = 0
        (0% loss),
    Approximate round trip times in milliseconds:
        Minimum = 16ms, Maximum = 16ms, average = 16ms
    Control-C
    ^C
```

j) Echo request to a specific server name.
```
    Z:\>PING SVRPRINC00

    Pinging SVRPRINC00 [192.168.5.240] with 32 bytes of data:
    Reply from 192.168.5.240: bytes = 32 time <1ms TTL = 128
    Reply from 192.168.5.240: bytes = 32 time <1ms TTL = 128
    Reply from 192.168.5.240: bytes = 32 time <1ms TTL = 128
    Reply from 192.168.5.240: bytes = 32 time <1ms TTL = 128

    Ping statistics for 192.168.5.240:
        Packets: Sent = 4, Received = 4, Lost = 0
        (0% loss),
    Approximate round trip times in milliseconds:
        Minimum = 0ms, Maximum = 0ms, Average = 0ms
```

k) The packet fragmentation no echo is established in the transmission.
```
    Z:\>PING  -f -a  192.168.5.240

    Pinging SVRPRINC00 [192.168.5.240] with 32 bytes of data:
    Reply from 192.168.5.240: bytes = 32 time <1ms TTL = 128
    Reply from 192.168.5.240: bytes = 32 time <1ms TTL = 128
    Reply from 192.168.5.240: bytes = 32 time <1ms TTL = 128
    Reply from 192.168.5.240: bytes = 32 time <1ms TTL = 128

    Ping statistics for 192.168.5.240:
        Packets: Sent = 4, Received = 4, Lost = 0
        (0% loss),
    Approximate round trip times in milliseconds:
        Minimum = 0ms, Maximum = 0ms, Average = 0ms
```

*Valid value range is 1 to 255*

l) Shelf life of the packages.
```
    Z:\>PING  -I 5  8.8.8.8

    Pinging 8.8.8.8 with 32 bytes of data:
    Request timed out.
    Request timed out.
    Request timed out.
    Request timed out.
```

```
    Ping statistics for 8.8.8.8:
        Packets: Sent = 4, Received = 0, Lost = 4
        (100% loss),

    Z:\>ping  -i 7  8.8.8.8

    Pinging 8.8.8.8 with 32 bytes of data:
    Request timed out.
    Request timed out.
    Request timed out.
    Request timed out.

    Ping statistics for 8.8.8.8:
        Packets: Sent = 4, Received = 0, Lost = 4
        (100% loss),

    Z:\>PING   -i 10   8.8.8.8

    Pinging 8.8.8.8 with 32 bytes of data:
    Reply from 216.239.48.249: TTL expired in transit.
    Reply from 216.239.48.249: TTL expired in transit.
    Reply from 216.239.48.249: TTL expired in transit.
    Reply from 216.239.48.249: TTL expired in transit.

    Ping statistics for 8.8.8.8:
        Packets: Sent = 4, Received = 4, Lost = 0
        (0% loss),

    Z:\>PING   -i 9   8.8.8.8

    Pinging 8.8.8.8 with 32 bytes of data:
    Reply from 72.14.234.231: TTL expired in transit.
    Reply from 72.14.234.231: TTL expired in transit.
    Reply from 72.14.234.231: TTL expired in transit.
    Reply from 72.14.234.231: TTL expired in transit.

    Ping statistics for 8.8.8.8:
        Packets: Sent = 4, Received = 4, Lost = 0
        (0% loss),

    Z:\>PING   -i 8   8.8.8.8

    Pinging 8.8.8.8 with 32 bytes of data:
    Response from 5.53.1.74: TTL expired in transit.
    Response from 5.53.1.74: TTL expired in transit.
    Response from 5.53.1.74: TTL expired in transit.
    Response from 5.53.1.74: TTL expired in transit.

    Ping statistics for 8.8.8.8:
        Packets: Sent = 4, Received = 4, Lost = 0
        (0% loss),

    Z:\>PING   -i 88   8.8.8.8

     Pinging 8.8.8.8 with 32 bytes of data:
    Reply from 8.8.8.8: bytes = 32 time = 16ms TTL = 51
    Reply from 8.8.8.8: bytes = 32 time = 16ms TTL = 51
    Reply from 8.8.8.8: bytes = 32 time = 16ms TTL = 51
    Reply from 8.8.8.8: bytes = 32 time = 16ms TTL = 51

    Ping statistics for 8.8.8.8:
        Packets: Sent = 4, Received = 4, Lost = 0
        (0% loss),
    Approximate round trip times in milliseconds:
        Minimum = 16ms, Maximum = 16ms,  average = 16ms
```

m) Check the registry path count jumps (only IPv4).
```
    Z:\>PING  -r 15   8.8.8.8
    Incorrect value for the -r option, the valid range is 1 to 9.

    Z:\>PING  -r 9   8.8.8.8

    Pinging 8.8.8.8 with 32 bytes of data:
    Request timed out.
    Request timed out.
    Request timed out.
    Request timed out.

    Ping statistics for 8.8.8.8:
        Packets: Sent = 4, Received = 0, Lost = 4
        (100% loss),
```

> Value or range for valid account hop route is between 1 and 9.

n)  Send buffer size.
```
Z:\>PING  -l 512   8.8.8.8

Pinging 8.8.8.8 with 512 bytes of data:
Reply from 8.8.8.8: bytes = 512 time = 18ms TTL = 51
Reply from 8.8.8.8: bytes = 512 time = 16ms TTL = 51
Reply from 8.8.8.8: bytes = 512 time = 16ms TTL = 51
Reply from 8.8.8.8: bytes = 512 time = 16ms TTL = 51

Ping statistics for 8.8.8.8:
    Packets: Sent = 4, Received = 4, Lost = 0
    (0% loss),
Approximate round trip times in milliseconds:
    Minimum = 16ms, Maximum = 18ms, average = 16ms
```

o)  Set the time count jumps (only IPv4).
```
Z:\>PING  -s 4   8.8.8.8

Pinging 8.8.8.8 with 32 bytes of data:
Request timed out.
Request timed out.
Request timed out.
Request timed out.

Ping statistics for 8.8.8.8:
    Packets: Sent = 4, Received = 0, Lost = 4
    (100% loss),

Z:\>PING  -s 4   -i 51 8.8.8.8

Pinging 8.8.8.8 with 32 bytes of data:
Request timed out.
Request timed out.
Request timed out.
Request timed out.

Ping statistics for 8.8.8.8:
    Packets: Sent = 4, Received = 0, Lost = 4
    (100% loss),
```

p)  Set the timeout in milliseconds for each response.
```
Z:\>PING  -w 5 8.8.8.8

Pinging 8.8.8.8 with 32 bytes of data:
Reply from 8.8.8.8: bytes = 32 time = 16ms TTL = 51
Reply from 8.8.8.8: bytes = 32 time = 16ms TTL = 51
Reply from 8.8.8.8: bytes = 32 time = 16ms TTL = 51
Reply from 8.8.8.8: bytes = 32 time = 16ms TTL = 51

Ping statistics for 8.8.8.8:
    Packets: Sent = 4, Received = 4, Lost = 0
    (0% loss),
Approximate round trip times in milliseconds:
    Minimum = 16ms, Maximum = 16ms, average = 16ms
```

p.1)  Time is 3 milliseconds.
```
Z:\>PING  -w 3 8.8.8.8

Pinging 8.8.8.8 with 32 bytes of data:
Reply from 8.8.8.8: bytes = 32 time = 16ms TTL = 51
Reply from 8.8.8.8: bytes = 32 time = 16ms TTL = 51
Reply from 8.8.8.8: bytes = 32 time = 16ms TTL = 51
Reply from 8.8.8.8: bytes = 32 time = 16ms TTL = 51

Ping statistics for 8.8.8.8:
    Packets: Sent = 4, Received = 4, Lost = 0
    (0% loss),
Approximate round trip times in milliseconds:
    Minimum = 16ms, Maximum = 16ms, average = 16ms
```

p.2)  Time is 2 milliseconds.
```
Z:\>PING  -w 2 8.8.8.8

Pinging 8.8.8.8 with 32 bytes of data:
Reply from 8.8.8.8: bytes = 32 time = 16ms TTL = 51
Reply from 8.8.8.8: bytes = 32 time = 16ms TTL = 51
Reply from 8.8.8.8: bytes = 32 time = 16ms TTL = 51
Reply from 8.8.8.8: bytes = 32 time = 16ms TTL = 51

Ping statistics for 8.8.8.8:
    Packets: Sent = 4, Received = 4, Lost = 0
    (0% loss),
Approximate round trip times in milliseconds:
```

                    Minimum = 16ms, Maximum = 16ms, average = 16ms
p.3) Time is 1 milliseconds.
        Z:\>PING   -w 1 8.8.8.8

        Pinging 8.8.8.8 with 32 bytes of data:
        Reply from 8.8.8.8: bytes = 32 time = 16ms TTL = 51
        Reply from 8.8.8.8: bytes = 32 time = 16ms TTL = 51
        Reply from 8.8.8.8: bytes = 32 time = 16ms TTL = 51
        Reply from 8.8.8.8: bytes = 32 time = 16ms TTL = 51

        Ping statistics for 8.8.8.8:
            Packets: Sent = 4, Received = 4, Lost = 0
            (0% loss),
        Approximate round trip times in milliseconds:
            Minimum = 16ms, Maximum = 16ms, average = 16ms
p.4) Time is 20 milliseconds.
        Z:\>PING   -w 20 8.8.8.8

        Pinging 8.8.8.8 with 32 bytes of data:
        Request timed out.
        Reply from 8.8.8.8: bytes = 32 time = 16ms TTL = 51
        Reply from 8.8.8.8: bytes = 32 time = 16ms TTL = 51
        Reply from 8.8.8.8: bytes = 32 time = 16ms TTL = 51

        Ping statistics for 8.8.8.8:
            Packets: Sent = 4, Received = 3 Lost = 1
            (25% loss),
        Approximate round trip times in milliseconds:
            Minimum = 16ms, Maximum = 16ms, average = 16ms
p.5) Time is 15 milliseconds.

        C:\Windows\system32>PING -w 15 8.8.8.8

        Pinging 8.8.8.8 with 32 bytes of data:
        Reply from 8.8.8.8: bytes = 32 time = 14ms TTL = 57
        Reply from 8.8.8.8: bytes = 32 time = 15ms TTL = 57
        Reply from 8.8.8.8: bytes = 32 time = 15ms TTL = 57
        Reply from 8.8.8.8: bytes = 32 time = 15ms TTL = 57

        Ping statistics for 8.8.8.8:
            Packets: Sent = 4, Received = 4, Lost = 0
            (0% loss),
        Approximate round trip times in milliseconds:
            Minimum = 14ms, Maximum = 15ms, average = 14ms
p.6) Time is 118 milliseconds.
        C:\Windows\system32>PING -w 118 8.8.8.8

        Pinging 8.8.8.8 with 32 bytes of data:
        Reply from 8.8.8.8: bytes = 32 time = 15ms TTL = 57
        Reply from 8.8.8.8: bytes = 32 time = 15ms TTL = 57
        Reply from 8.8.8.8: bytes = 32 time = 18ms TTL = 57
        Reply from 8.8.8.8: bytes = 32 time = 15ms TTL = 57

        Ping statistics for 8.8.8.8:
            Packets: Sent = 4, Received = 4, Lost = 0
            (0% loss),
        Approximate round trip times in milliseconds:
            Minimum = 15ms, Maximum = 18ms, average =  15ms

q)  Set the maximum number of hops (along the ROUTERS).
        Z:\>ping -S 192.168.10.5    8.8.8.8

        Pinging from 192.168.10.5 8.8.8.8 with 32 bytes of data:
        PING: error in transmission. General error.
        PING: error in transmission. General error.
        PING: error in transmission. General error.
        PING: error in transmission. General error.

        Ping statistics for 8.8.8.8:
            Packets: Sent = 4, Received = 0, Lost = 4
            (100% loss),

        Z:\>ping -S localhost    8.8.8.8

        Pinging 8.8.8.8 from 127.0.0.1 with 32 bytes of data:
        PING: error in transmission. General error.
        PING: error in transmission. General error.
        PING: error in transmission. General error.
        PING: error in transmission. General error.

        Ping statistics for 8.8.8.8:

> Card address or localhost: 127.0.0.1

```
            Packets: Sent = 4, Received = 0, Lost = 4
            (100% loss),
```

## STEP 4.2: PING executed from SERVER

a)  It is checked on a Windows Server 2012 R2 Server, the effect of the same command on a Windows 10 client.
```
    C:\Windows\system32>ping -j  8.8.8.8

    Pinging 8.8.8.8 with 32 bytes of data:
    General error.
    General error.
    General error.
    General error.

    Ping statistics for 8.8.8.8:
        Packets: Sent = 4, Received = 0, Lost = 4
        (100% loss),
```

b)  Normal ping IP
```
    C:\Windows\system32>ping 8.8.8.8

    Pinging 8.8.8.8 with 32 bytes of data:
    Reply from 8.8.8.8: bytes = 32 time = 16ms TTL = 51
    Reply from 8.8.8.8: bytes = 32 time = 16ms TTL = 51
    Reply from 8.8.8.8: bytes = 32 time = 16ms TTL = 51
    Reply from 8.8.8.8: bytes = 32 time = 16ms TTL = 51

    Ping statistics for 8.8.8.8:
        Packets: Sent = 4, Received = 4, Lost = 0
        (0% loss),
    Approximate round trip times in milliseconds:
        Minimum = 16ms, Maximum = 16ms, average = 16ms
```

c)  Strict source route for list-IP host and a mandatory output is set.
```
    C:\Windows\system32>ping -j 192.168.5.240  -S 127.0.0.1  8.8.8.8

    Pinging 8.8.8.8 from 127.0.0.1 with 32 bytes of data:
    PING: error in transmission. General error.
    PING: error in transmission. General error.
    PING: error in transmission. General error.
    PING: error in transmission. General error.

    Ping statistics for 8.8.8.8:
        Packets: Sent = 4, Received = 0, Lost = 4
        (100% loss),
```

d)  Establish ping using IPv6, from two different servers have two different cards (:: 2004 :: 200e).
```
    C:\Windows\system32>ping -6 google.es

    Pinging google.es [2a00: 1450: 4003: 2003 :: 805] with 32 bytes of data:
    Request timed out.
    Request timed out.
    Request timed out.
    Request timed out.

    Ping statistics for 2a00: 1450: 4003: 2003 :: 805:
        Packets: Sent = 4, Received = 0, Lost = 4
        (100% loss),

    C:\Windows\system32>ping -6 google.com

    Pinging google.com [2a00: 1450: 4003: 805 :: 200e] with 32 bytes of data:
    Request timed out.
    Request timed out.
    Request timed out.
    Request timed out.

    Ping statistics for 2a00: 1450: 4003: 805 :: 200e:
        Packets: Sent = 4, Received = 0, Lost = 4
        (100% loss),
```

## STEP 5: Setting and display of IP addresses. IPCONFIG

Command or console application that displays the network settings of TCP / IP current and updated or rege-nera the DHCP protocol and domain name system (DNS). There have been tools with graphical interface called winipcfg (Windows 98) and wntipcfg (Windows NT).
    IPCONFIG

a)  Display default values of the network card.
```
    C:\Windows\system32>ipconfig

    Windows IP Configuration
```

```
Ethernet adapter Ethernet:

   Connection-specific DNS Suffix  . :
   IPv4 Address. . . . . . . . . . . : 192.168.1.89
   Subnet Mask . . . . . . . . . . . : 255.255.255.0
   IPv4 Address. . . . . . . . . . . : 192.168.2.89
   Subnet Mask . . . . . . . . . . . : 255.255.255.0
   IPv4 Address. . . . . . . . . . . : 192.168.5.240
   Subnet Mask . . . . . . . . . . . : 255.255.255.0
   Default Gateway . . . . . . . . . : 192.168.2.100
                                       192.168.1.1

Tunnel adapter isatap.{0D7F8ED4-1EC1-4BB3-A058-D0C317F404A5}:

   Media State . . . . . . . . . . . : Media disconnected
   Connection-specific DNS Suffix  . :
```

> **DHCP Enabled** Indicates whether DHCP service is enabled or not. Autoconfiguration enabled Indicates if we have our network settings automatically.
> **Link:** Local IPv6 address: our sample IPv6 address of your machine (SO that support it).
> **IPv4 Address:** displays the current IP address of your machine.
> **Subnet Mask:** shows what subnet mask our network.
> **Default Gateway** Displays the IP gateway (usually our router).
> **DHCP** server displays the IP from the DHCP server to which we are connected.
> **IAID DHCPv6:** shows information about DHCP in IPv6 version (SO that support it).
> **DNS servers:** shows the IP of the DNS servers to which we are connected.

b) Display all information of the network card.
```
Z:\> IPCONFIG   /ALL
Windows IP Configuration

   Host Name . . . . . . . . . . . . : SVRPRINC00
   Primary Dns Suffix  . . . . . . . : bspWeb.local
   Node Type . . . . . . . . . . . . : Hybrid
   IP Routing Enabled. . . . . . . . : No
   WINS Proxy Enabled. . . . . . . . : No
   DNS Suffix Search List. . . . . . : bspWeb.local

Ethernet adapter Ethernet:

   Connection-specific DNS Suffix  . :
   Description . . . . . . . . . . . : Microsoft Hyper-V Network Adapter
   Physical Address. . . . . . . . . : 00-15-5D-01-63-06
   DHCP Enabled. . . . . . . . . . . : No
   Autoconfiguration Enabled . . . . : Yes
   IPv4 Address. . . . . . . . . . . : 192.168.1.89(Preferred)
   Subnet Mask . . . . . . . . . . . : 255.255.255.0
   IPv4 Address. . . . . . . . . . . : 192.168.2.89(Preferred)
   Subnet Mask . . . . . . . . . . . : 255.255.255.0
   IPv4 Address. . . . . . . . . . . : 192.168.5.240(Preferred)
   Subnet Mask . . . . . . . . . . . : 255.255.255.0
   Default Gateway . . . . . . . . . : 192.168.2.100
                                       192.168.1.1
   DNS Servers . . . . . . . . . . . : 127.0.0.1
   NetBIOS over Tcpip. . . . . . . . : Enabled

Tunnel adapter isatap.{0D7F8ED4-1EC1-4BB3-A058-D0C317F404A5}:

   Media State . . . . . . . . . . . : Media disconnected
   Connection-specific DNS Suffix  . :
   Description . . . . . . . . . . . : Microsoft ISATAP Adapter #2
   Physical Address. . . . . . . . . : 00-00-00-00-00-00-00-E0
   DHCP Enabled. . . . . . . . . . . : No
   Autoconfiguration Enabled . . . . : Yes
```
c) To renew the network adapters (DHCP) Assign a new IP address.
   IPCONFIG  /RENEW
   IPCONFIG  /RENEW6
d) Display the DNS servers (used and cached).
```
Z:\> IPCONFIG   /DISPLAYDNS
Windows IP Configuration

c.autoscout24.com
----------------------------------------
Record Name . . . . . : c.autoscout24.com
Record Type . . . . . : 1
Time To Live  . . . . : 143
Data Length . . . . . : 4
Section . . . . . . . : Answer
A (Host) Record . . . : 92.42.227.196

_ldap._tcp.svrprinc00.bspweb.local
----------------------------------------
Name does not exist.

ww3.autoscout24.es
----------------------------------------
. . . . . . . . . . . . . .
```

e) Check if there are any DNS, DHCP server.

        IPCONFIG  /registerDNS
```
Windows IP Configuration

registration of DNS resource records for all adapters of this equipment began. Any error is reported
in the Event Viewer in 15 minutes.
```
f) Display all IP addresses available on a DHCP server.
        IPCONFIG  /SHOWCLASSID
        IPCONFIG  /SHOWCLASSID6
g) Refresh Cache.
        IPCONFIG  /FLUSHDNS
h) Release IPS addresses.
        IPCONFIG  /RELEASE
        IPCONFIG  /RELEASE6
i) Shared resources. Display information of the compartments..
        IPCONFIG  /ALLCOMPARTMENTS
```
C:\Windows\system32>IPCONFIG    /ALLCOMPARTMENTS

Windows IP Configuration

==============================================================================
Network Information for Compartment 1 (ACTIVE)
==============================================================================

Ethernet adapter Ethernet:

   Connection-specific DNS Suffix  . :
   IPv4 Address. . . . . . . . . . . : 192.168.1.89
   Subnet Mask . . . . . . . . . . . : 255.255.255.0
   IPv4 Address. . . . . . . . . . . : 192.168.2.89
   Subnet Mask . . . . . . . . . . . : 255.255.255.0
   IPv4 Address. . . . . . . . . . . : 192.168.5.240
   Subnet Mask . . . . . . . . . . . : 255.255.255.0
   Default Gateway . . . . . . . . . : 192.168.2.100
                                       192.168.1.1

Tunnel adapter isatap.{0D7F8ED4-1EC1-4BB3-A058-D0C317F404A5}:

   Media State . . . . . . . . . . . : Media disconnected
   Connection-specific DNS Suffix  . :

C:\Windows\system32>IPCONFIG    /ALLCOMPARTMENTS /all

Windows IP Configuration

==============================================================================
Network Information for Compartment 1 (ACTIVE)
==============================================================================

   Host Name . . . . . . . . . . . . : SVRPRINC00
   Primary Dns Suffix  . . . . . . . : bspWeb.local
   Node Type . . . . . . . . . . . . : Hybrid
   IP Routing Enabled. . . . . . . . : No
   WINS Proxy Enabled. . . . . . . . : No
   DNS Suffix Search List. . . . . . : bspWeb.local

Ethernet adapter Ethernet:

   Connection-specific DNS Suffix  . :
   Description . . . . . . . . . . . : Microsoft Hyper-V Network Adapter
   Physical Address. . . . . . . . . : 00-15-5D-01-63-06
   DHCP Enabled. . . . . . . . . . . : No
   Autoconfiguration Enabled . . . . : Yes
   IPv4 Address. . . . . . . . . . . : 192.168.1.89(Preferred)
   Subnet Mask . . . . . . . . . . . : 255.255.255.0
   IPv4 Address. . . . . . . . . . . : 192.168.2.89(Preferred)
   Subnet Mask . . . . . . . . . . . : 255.255.255.0
   IPv4 Address. . . . . . . . . . . : 192.168.5.240(Preferred)
   Subnet Mask . . . . . . . . . . . : 255.255.255.0
   Default Gateway . . . . . . . . . : 192.168.2.100
                                       192.168.1.1
   DNS Servers . . . . . . . . . . . : 127.0.0.1
   NetBIOS over Tcpip. . . . . . . . : Enabled

Tunnel adapter isatap.{0D7F8ED4-1EC1-4BB3-A058-D0C317F404A5}:

   Media State . . . . . . . . . . . : Media disconnected
   Connection-specific DNS Suffix  . :
   Description . . . . . . . . . . . : Microsoft ISATAP Adapter #2
   Physical Address. . . . . . . . . : 00-00-00-00-00-00-00-E0
   DHCP Enabled. . . . . . . . . . . : No
```

```
        Autoconfiguration Enabled . . . . : Yes
```

j) Modify the IP address assigned by DHCP SERVER respect to the object.
   IPCONFIG /SETCLASSID
   IPCONFIG /SETCLASSID6
k) Display adapter information and DHCP for IPv4.
```
   Z:\> IPCONFIG    /SHOWCLASSID    *

   Windows IP Configuration

   There are no DHCPv4 classes defined for Ethernet.
   Unable to modify the DHCPv4 class id for adapter Loopback Pseudo-Interface 1: The
   system cannot find the file specified.
```
l) Display adapter information and .. dhcp for IPv6.
   IPCONFIG /SHOWCLASSID6 *
m) Display all information of components.
   IPCONFIG /ALLCOMPARTMENTS
   IPCONFIG /ALLCOMPARTMENTS /ALL
```
   Z:\> IPCONFIG    /ALLCOMPARTMENTS    /ALL
   C:\Windows\system32>IPCONFIG    /ALLCOMPARTMENTS /all
   Windows IP Configuration


   ==========================================================================
   Network Information for Compartment 1 (ACTIVE)
   ==========================================================================
        Host Name . . . . . . . . . . . . : SVRPRINC00
        Primary Dns Suffix  . . . . . . . : bspWeb.local
        Node Type . . . . . . . . . . . . : Hybrid
        IP Routing Enabled. . . . . . . . : No
        WINS Proxy Enabled. . . . . . . . : No
        DNS Suffix Search List. . . . . . : bspWeb.local

   Ethernet adapter Ethernet:

        Connection-specific DNS Suffix  . :
        Description . . . . . . . . . . . : Microsoft Hyper-V Network Adapter
        Physical Address. . . . . . . . . : 00-15-5D-01-63-06
        DHCP Enabled. . . . . . . . . . . : No
        Autoconfiguration Enabled . . . . : Yes
        IPv4 Address. . . . . . . . . . . : 192.168.1.89(Preferred)
        Subnet Mask . . . . . . . . . . . : 255.255.255.0
        IPv4 Address. . . . . . . . . . . : 192.168.2.89(Preferred)
        Subnet Mask . . . . . . . . . . . : 255.255.255.0
        IPv4 Address. . . . . . . . . . . : 192.168.5.240(Preferred)
        Subnet Mask . . . . . . . . . . . : 255.255.255.0
        Default Gateway . . . . . . . . . : 192.168.2.100
                                            192.168.1.1
        DNS Servers . . . . . . . . . . . : 127.0.0.1
        NetBIOS over Tcpip. . . . . . . . : Enabled

   Tunnel adapter isatap.{0D7F8ED4-1EC1-4BB3-A058-D0C317F404A5}:

        Media State . . . . . . . . . . . : Media disconnected
        Connection-specific DNS Suffix  . :
        Description . . . . . . . . . . . : Microsoft ISATAP Adapter #2
        Physical Address. . . . . . . . . : 00-00-00-00-00-00-00-E0
        DHCP Enabled. . . . . . . . . . . : No
        Autoconfiguration Enabled . . . . : Yes
```

## STEP 6: Display full information on the network card. NIC from WMIC.

WMIC (Windows Management Instrumentation Command-line), is a management tool for Windows that allows not only information but perform actions.
- Within each alias we can find several values and which we can take action.
- We can serve for system values.
- You can change attribute values of certain objects.

a) Using the WMI application. Windows Manager Interface Console.
   C:\> WMIC
   Access the WMI application, and your prompt disappears.
   wmic:root\cli>
b) Get the help of WMIC commands within the application.
   wmic:root\cli>  /?
c) Display information network cards. NIC appears without ordering information.
   wmic:root\cli> NIC

d) Display list, all information network card NIC.
   wmic:root\cli> NIC LIST FULL
e) Display the same information pto. d) running it from the command line.
   C:\> WMIC NIC LIST FULL
f) Perform a flush DNS.
   C:\> WMIC NICCONFIG CALL FLUSHDNS
g) Check the serial number of the BIOS of a Computer.
   C:\> WMIC BIOS GET SERIALNUMBER
h) Display System shares in table format.
```
C:\> WMIC SHARE LIST /FORMAT:TABLE
AccessMask  AllowMaximum  Description       InstallDate  MaximumAllowed  Name    Pa
th          Status        Type
            TRUE                Remote Admin                             ADMIN$  C:
\Windows    OK            2147483648
            TRUE                Default share                            C$      C:
\           OK            2147483648
            TRUE                Remote IPC                               IPC$
            OK            2147483651
```

## STEP 7: Identification team, SID, FQDN. WHOAMI

This utility can be used to obtain the destination information together with the respective identifiers Safety-ness (SID), notifications, privileges, login ID.

a) Display domain information.
   WHOAMI
b) Display default.
   WHOAMI
   bspweb\administrador
c) Display the main user name (UPN).
   WHOAMI /UPN
   administrador@bspweb.local
d) Display the fully qualified name.
```
Z:\> WHOAMI   /FQDN
CN=Administrador,CN=Users,DC=bspWeb,DC=local
```
e) Display domain users.
```
C:\Windows\system32>whoami /user
USER INFORMATION
----------------

User Name                SID
======================== =============================================
bspweb0\administrator    S-1-5-21-2214572230-4234099076-1720029264-500

C:\Windows\system32>wHOAMI   /USER   /FO TABLE
USER INFORMATION
----------------

User Name                SID
======================== =============================================
bspweb0\administrator    S-1-5-21-2214572230-4234099076-1720029264-500
```
f) Display domain users with delimiter CSV format (in quotes and comma delimited).
   C:\Windows\system32>WHOAMI   /USER   /FO   CSV
   "Nombre de usuario","SID"
   "bspweb\administrador","S-1-5-21-2298800814-2528216055-1890261488-500"
g) Display domain users with LIST format.
```
C:\Windows\system32>WHOAMI   /USER   /FO  LIST
USER INFORMATION
----------------

User Name                        SID
================================ =============================================
svr-prin-001\administrator       S-1-5-21-2214572230-4234099076-1720029264-500
```
h) Display group information.
   WHOAMI /GROUPS
   WHOAMI /GROUPS      /FO CSV
   WHOAMI /GROUPS      /FO LIST
   WHOAMI /GROUPS      /FO TABLE
```
C:\Windows\system32>WHOAMI    /GROUPS  /FO    TABLE

GROUP INFORMATION
-----------------
```

```
Group Name                                              Type             SID
                                 Attributes
======================================== ================ ================
=============
Everyone                                                Well-known group S-1-1-0
                                 Mandatory group, Enabled by default, Enabled group
BUILTIN\Administrators                                  Alias            S-1-5-32-544
                                 Mandatory group, Enabled by default, Enabled group
, Group owner
BUILTIN\Remote Desktop Users                            Alias            S-1-5-32-555
                                 Mandatory group, Enabled by default, Enabled group
BUILTIN\Users                                           Alias            S-1-5-32-545
                                 Mandatory group, Enabled by default, Enabled group
BUILTIN\Pre-Windows 2000 Compatible Access              Alias            S-1-5-32-554
                                 Mandatory group, Enabled by default, Enabled group
NT AUTHORITY\REMOTE INTERACTIVE LOGON                   Well-known group S-1-5-14
                                 Mandatory group, Enabled by default, Enabled group
NT AUTHORITY\INTERACTIVE                                Well-known group S-1-5-4
                                 Mandatory group, Enabled by default, Enabled group
NT AUTHORITY\Authenticated Users                        Well-known group S-1-5-11
                                 Mandatory group, Enabled by default, Enabled group
NT AUTHORITY\This Organization                          Well-known group S-1-5-15
                                 Mandatory group, Enabled by default, Enabled group
LOCAL                                                   Well-known group S-1-2-0
                                 Mandatory group, Enabled by default, Enabled group
BSPWEB0\Domain Admins                                   Group            S-1-5-21-2214572
230-4234099076-1720029264-512 Mandatory group, Enabled by default, Enabled group
BSPWEB0\Group Policy Creator Owners                     Group            S-1-5-21-2214572
230-4234099076-1720029264-520 Mandatory group, Enabled by default, Enabled group
BSPWEB0\Enterprise Admins                               Group            S-1-5-21-2214572
230-4234099076-1720029264-519 Mandatory group, Enabled by default, Enabled group
BSPWEB0\Schema Admins                                   Group            S-1-5-21-2214572
230-4234099076-1720029264-518 Mandatory group, Enabled by default, Enabled group
Authentication authority asserted identity              Well-known group S-1-18-1
                                 Mandatory group, Enabled by default, Enabled group
BSPWEB0\Denied RODC Password Replication Group Alias             S-1-5-21-2214572
230-4234099076-1720029264-572 Mandatory group, Enabled by default, Enabled group
, Local Group
```

i) `Mandatory Label\High Mandatory Level          Label            S-1-16-12288` Notifications -> KERBEROS.

WHOAMI   /CLAIMS

```
C:\Windows\system32>WHOAMI        /CLAIMS

USER CLAIMS INFORMATION
-----------------------

User claims unknown.

Kerberos support for Dynamic Access Control on this device has been disabled.
```

j) Display privileges, description and status (objects).

WHOAMI   /PRIV

```
C:\Windows\system32>WHOAMI        /PRIV

PRIVILEGES INFORMATION
----------------------

Privilege Name                    Description                                    State
================================= ============================================== ========
SeIncreaseQuotaPrivilege          Adjust memory quotas for a process             Disabled
SeMachineAccountPrivilege         Add workstations to domain                     Disabled
SeSecurityPrivilege               Manage auditing and security log               Disabled
SeTakeOwnershipPrivilege          Take ownership of files or other objects       Disabled
SeLoadDriverPrivilege             Load and unload device drivers                 Disabled
SeSystemProfilePrivilege          Profile system performance                     Disabled
SeSystemtimePrivilege             Change the system time                         Disabled
SeProfileSingleProcessPrivilege   Profile single process                         Disabled
SeIncreaseBasePriorityPrivilege   Increase scheduling priority                   Disabled
SeCreatePagefilePrivilege         Create a pagefile                              Disabled
SeBackupPrivilege                 Back up files and directories                  Disabled
SeRestorePrivilege                Restore files and directories                  Disabled
SeShutdownPrivilege               Shut down the system                           Disabled
SeDebugPrivilege                  Debug programs                                 Disabled
SeSystemEnvironmentPrivilege      Modify firmware environment values             Disabled
SeChangeNotifyPrivilege           Bypass traverse checking                       Enabled
SeRemoteShutdownPrivilege         Force shutdown from a remote system            Disabled
SeUndockPrivilege                 Remove computer from docking station           Disabled
SeEnableDelegationPrivilege       Enable computer and user accounts to be trusted
for delegation                                                                   Disabled
SeManageVolumePrivilege           Perform volume maintenance tasks               Disabled
SeImpersonatePrivilege            Impersonate a client after authentication      Enabled
```

```
SeCreateGlobalPrivilege          Create global objects                       Enabled
SeIncreaseWorkingSetPrivilege    Increase a process working set              Disabled
SeTimeZonePrivilege              Change the time zone                        Disabled
SeCreateSymbolicLinkPrivilege    Create symbolic links                       Disabled
```

k) Display current connection information.

    WHOAMI /LOGONID

```
Z:\> WHOAMI  /LOGONID
S-1-5-5-0-177112
```

l) Display all information.

    WHOAMI /ALL

```
Z:\> WHOAMI   /ALL

USER INFORMATION
----------------

User Name                    SID
=========================    ============================================
svr-prin-001\administrator   S-1-5-21-2214572230-4234099076-1720029264-500


GROUP INFORMATION
-----------------

Group Name                                                    Type             SID              At-
tributes
============================================================  ================ ============
Everyone                                                      Well-known group S-1-1-0
Mandatory group, Enabled by default, Enabled group
NT AUTHORITY\Local account and member of Administrators group Well-known group S-1-5-114
Mandatory group, Enabled by default, Enabled group
BUILTIN\Administrators                                        Alias            S-1-5-32-544
Mandatory group, Enabled by default, Enabled group, Group owner
BUILTIN\Users                                                 Alias            S-1-5-32-545
Mandatory group, Enabled by default, Enabled group
NT AUTHORITY\REMOTE INTERACTIVE LOGON                         Well-known group S-1-5-14
Mandatory group, Enabled by default, Enabled group
NT AUTHORITY\INTERACTIVE                                      Well-known group S-1-5-4
Mandatory group, Enabled by default, Enabled group
NT AUTHORITY\Authenticated Users                              Well-known group S-1-5-11
Mandatory group, Enabled by default, Enabled group
NT AUTHORITY\This Organization                                Well-known group S-1-5-15
Mandatory group, Enabled by default, Enabled group
NT AUTHORITY\Local account                                    Well-known group S-1-5-113
Mandatory group, Enabled by default, Enabled group
LOCAL                                                         Well-known group S-1-2-0
Mandatory group, Enabled by default, Enabled group
NT AUTHORITY\NTLM Authentication                              Well-known group S-1-5-64-10
Mandatory group, Enabled by default, Enabled group
Mandatory Label\High Mandatory Level                          Label            S-1-16-12288


PRIVILEGES INFORMATION
----------------------

Privilege Name                   Description                                 State
==============================   =========================================   ========
SeIncreaseQuotaPrivilege         Adjust memory quotas for a process          Disabled
SeSecurityPrivilege              Manage auditing and security log            Disabled
SeTakeOwnershipPrivilege         Take ownership of files or other objects    Disabled
SeLoadDriverPrivilege            Load and unload device drivers              Disabled
SeSystemProfilePrivilege         Profile system performance                  Disabled
SeSystemtimePrivilege            Change the system time                      Disabled
SeProfileSingleProcessPrivilege  Profile single process                      Disabled
SeIncreaseBasePriorityPrivilege  Increase scheduling priority                Disabled
SeCreatePagefilePrivilege        Create a pagefile                           Disabled
SeBackupPrivilege                Back up files and directories               Disabled
SeRestorePrivilege               Restore files and directories               Disabled
SeShutdownPrivilege              Shut down the system                        Disabled
SeDebugPrivilege                 Debug programs                              Disabled
SeSystemEnvironmentPrivilege     Modify firmware environment values          Disabled
SeChangeNotifyPrivilege          Bypass traverse checking                    Enabled
SeRemoteShutdownPrivilege        Force shutdown from a remote system         Disabled
SeUndockPrivilege                Remove computer from docking station        Disabled
SeManageVolumePrivilege          Perform volume maintenance tasks            Disabled
SeImpersonatePrivilege           Impersonate a client after authentication   Enabled
SeCreateGlobalPrivilege          Create global objects                       Enabled
SeIncreaseWorkingSetPrivilege    Increase a process working set              Disabled
SeTimeZonePrivilege              Change the time zone                        Disabled
SeCreateSymbolicLinkPrivilege    Create symbolic links                       Disabled
```

# PRACTICE 2: Configure Network structure: UO, groups and users.

## DESCRIPTION:

### What is an Organizational Unit (OU)?

OU is an object type very useful directory included in domains. Organizational units are Active Directory containers into which you can place users, groups, computers, and other organizational units. An organizational unit can not contain objects from other domains.

An organizational unit is the smallest scope or unit to which you can assign Group Policy settings or you can delegate administrative authority. With OUs, you can create containers within a domain that represent the logical and existing hierarchical structures within an organization. This allows you to manage the configuration and use of accounts and resources based on your organizational model

### What is a Group AD DS?

A group is a set of user accounts and equipment, contacts and other groups that can be managed as a single unit. Users and computers that belong to a particular group are called members.

Groups Domain Services Active Directory (AD DS) are directory objects that reside in a domain and container objects organizational unit (OU). AD DS provides a set of default groups when installed and also includes an option to create them.

AD DS groups can be used to:
- Simplify administration by assigning permissions for a share to a group rather than individual users resource. When permissions are assigned to a group, the same access to the resource is granted to all members of that group.
- Delegate administration by assigning user rights to a group at once using Group Policy. Then, you can add that group members will want to have the same rights as the group.
- Create lists of email distribution.

The groups are characterized by their scope and type.
- The scope of a group determines the scope of the group within a domain or forest: Domain Local Groups, Global Groups and Universal Groups.
- The group type determines whether you can use a group to assign permissions from a shared resource (for security groups) or if you can use a single group distribution lists for email (for distribution groups).

There are also groups whose group memberships can not be viewed or modified. These groups are known by the nom-bre special identities. They represent different users at different times, depending on the circumstances. For example, the Everyone group is a special identity that represents all current network users, including guests and users from other domains.

### What is a User AD DS?

User accounts Active Directory represent physical entities, such as people. Also you can be used as dedicated service accounts for some applications.

Sometimes user accounts are also called security principals. Security principals are directory objects that automatically security identifiers (SIDs), which can be used to access domain resources are assigned. Primarily, a user account:

- **Authenticate the identity of a user:** A user account allows a user to log on to computers and domains with an identity that the domain can authenticate. A user who logs on to the network must have a user account and own unique password. To maximize security, avoid multiple users sharing one account.
- **Authorizes or denies access to domain resources**: After a user authenticates, it is granted or denied access to domain resources based on the explicit permissions that are assigned to you in the action.

## STEP 1: View shared drives NET USE

a) View shared drives.

NET USE
```
C:\Windows\SYSVOL\sysvol\bspWeb.local>NET USE
New connections are recorded.

No entries in the list.
```

> Disable shared resources quickly in Windows
> Access REGEDIT and search for:
> HKEY_LOCAL_MACHINE\System\CurrentControlSet\Services\LanmanServer\Parameters
> Once we get there we must create a DWORD value named parameters AutoShareWks and whose value is 0.

b) Display shared resources.

NET SHARE
```
C:\Windows\SYSVOL\sysvol\bspWeb.local>NET SHARE

Nombre          Recurso                                  Descripción
-------------------------------------------------------------------------------
C$              C:\                                      Recurso predeterminado
IPC$                                                     IPC remota
ADMIN$          C:\Windows                               Admin remotaCARPETADEUSUARIOS
                C:\Windows\SYSVOL\sysvol\bspWeb.local\repaso
NETLOGON        C:\Windows\SYSVOL\sysvol\bspWeb.local\SCRIPTS
                                                         Recurso compartido del servidor...
SYSVOL          C:\Windows\SYSVOL\sysvol                 Recurso compartido del servidor...
Se ha completado el comando correctamente.
```

## STEP 2: Share the resource NET USE

a) Share resource.
   SET LOGO
   ```
   C:\Windows\SYSVOL\sysvol\bspWeb.local>SET LOGO
   LOGONSERVER=\\SVRPRINC00
         NET  USE M:    %LOGONSERVER%\CARPETADEUSUARIOS
         NET  USE  M:    \\SVRPRINC00\CARPETADEUSUARIOS
   ```

b) Command NET.

b.1) Assign a shared resource to a drive.
   Create directory
      We see within the domain
      ```
      c:\windows\sysvol\sysvol\bspweb.local>MD    REPASO
      ```

> NOTE: A Resource hidden share: **name$**

> **Shared resources**
> Disk volume: C $, D $, ...
> Operating System folder: ADMIN$
> Cache Fax: Folder containing fax, FAX$.
> Communication: IPC$
> Printers folder: PRINT$
> Domain controllers: SYSVOL, NETLOGON not have $

b.2) Access the Windows graphical environment to share the resource. It is done in a graphical format as it is much easier to assign permissions comportación.

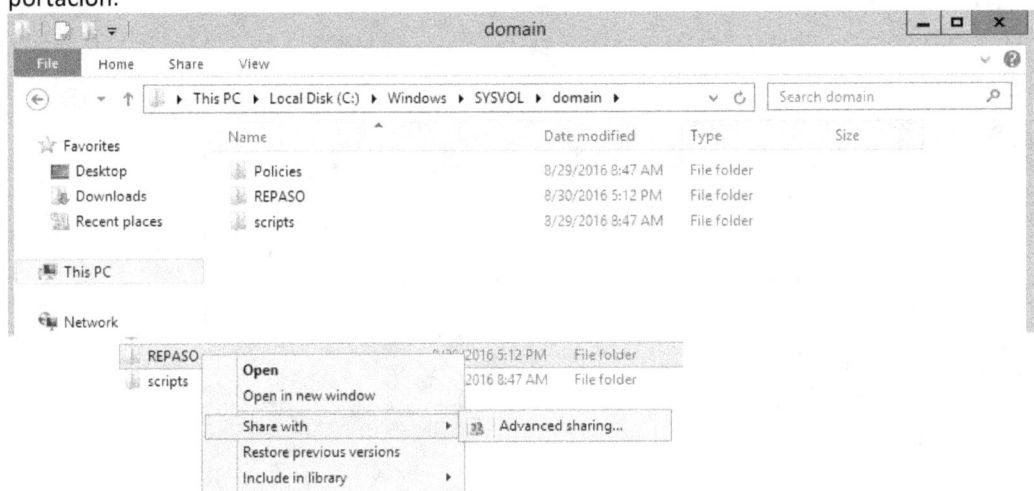

b.3) Access Regedit.
   C:\> REGEDIT
      Review.
         Right button.

> **Enable administrative shares in Windows 7**
> Access: REGEDIT
> HKEY_LOCAL_MACHINE\Software\Microsoft\Windows\CurrentVersion\Policies\System
> Create the name: **LocalAccountTokenFilterPolicy**
> Type: **DWORD (for 32 or 64 bits).**
> Data: **hexadecimal value of 1**

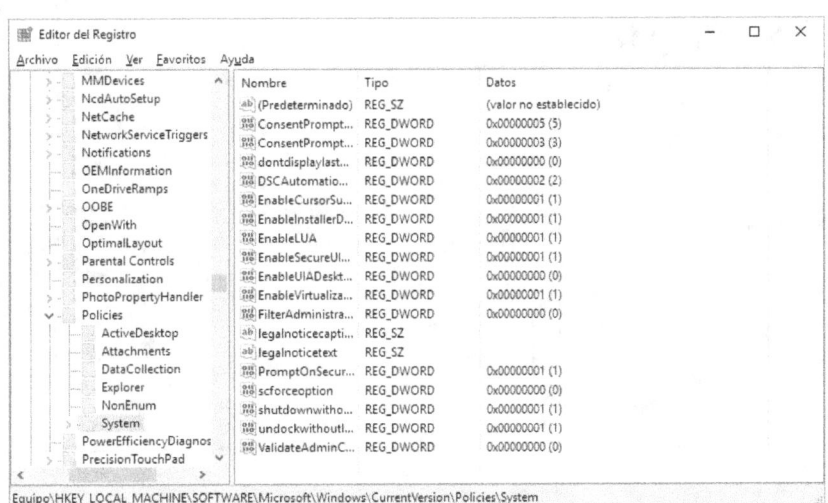

c) Properties on the share.
   Right button.

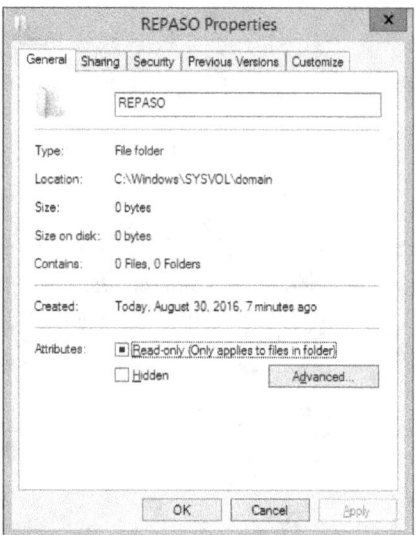

d) Lapel share.

    Select the Share button ...

        Permissions window appears shared resources

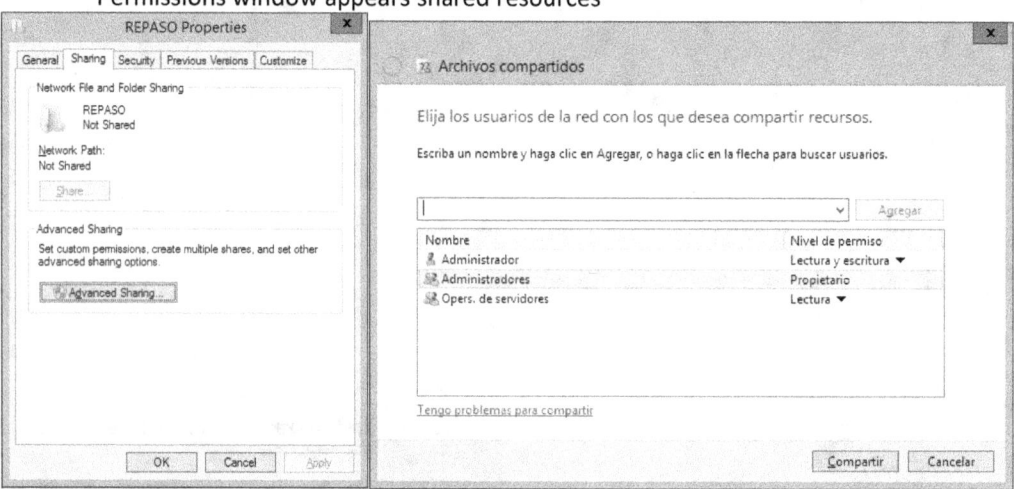

    Advanced Sharing button Use ... To set custom permissions and create multiple shared resources and set other advanced options for sharing

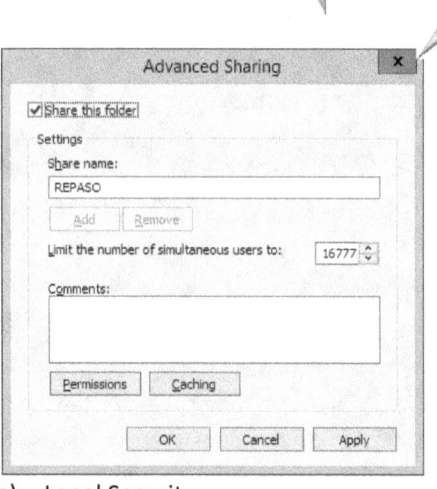

Default 167777

It allows you to set advanced permissions to the share.

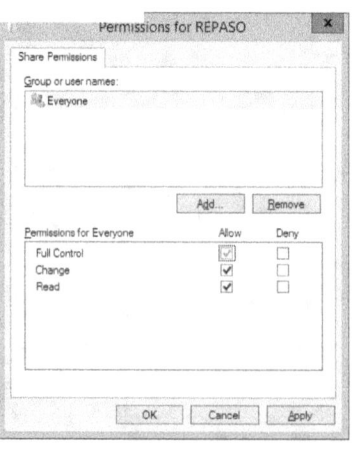

Add...
Add groups or users defined in the system.

e) Lapel Security.

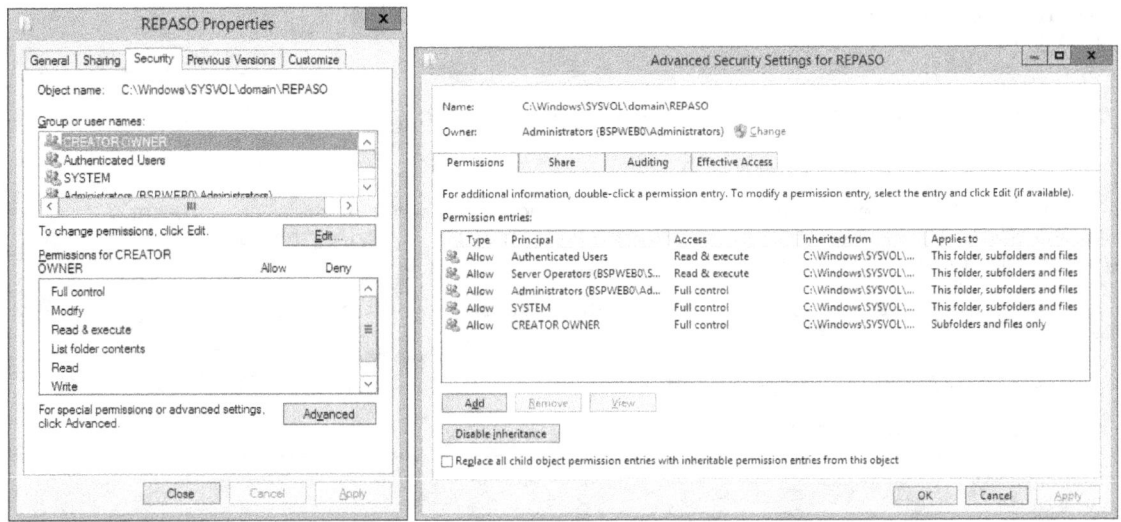

## STEP 2.1: Ask for shares NET SHARE
```
NET USE
NET SHARE
```
a) On a domain server. Assign a mount point to a UNIT
```
C:\Windows\SYSVOL\sysvol\bspWeb.local>NET USE M: \\SVRPRIN00\REPASO

C:\Windows\SYSVOL\sysvol\bspWeb.local>NET USE N: \\SVRPRIN00\CARPETAUSUARIO
```
b) Check the units assigned to shared resources.
```
C:\Windows\SYSVOL\sysvol\bspWeb.local>NET USE
New connections will be remembered.


Status       Local       Remote                     Network

-------------------------------------------------------------------------------
OK           M:          \\SVRPRINC00\REPASO        Microsoft Windows Network
The command completed successfully.
```
c) Display shared resources on a local computer.
```
C:\Windows\SYSVOL\sysvol\bspWeb.local>net share

Share name    Resource                                    Remark

-------------------------------------------------------------------------------
C$            C:\                                         Default share
IPC$                                                      Remote IPC
ADMIN$        C:\Windows                                  Remote Admin
CARPETAUSERIOS
              C:\Windows\SYSVOL\domain\REP...
NETLOGON      C:\Windows\SYSVOL\sysvol\bspWeb.local\SCRIPTS
                                                          Logon server share
REPASO        C:\Windows\SYSVOL\domain\REP...
SYSVOL        C:\Windows\SYSVOL\sysvol                    Logon server share
The command completed successfully.
```

> List the shares on a Windows server from Linux:
>   smbclient -L nombre_de_host -U%
> # smbclient -L \\SVRPRINC00
> # smbclient -L \\SVRPRINC00 -U%

## STEP 3: Create groups and consultations NET GROUP
a) Create groups and consult system groups.

a.1) Used only on a Windows Domain Controller.
```
C:\Users\Administrator>net group
This command can be used only on a Windows Domain Controller.

More help is available by typing NET HELPMSG 3515.
```
a.2) View groups in the operating system.
```
NET  GROUP
C:\Windows\SYSVOL\sysvol\bspWeb.local>net group
Group Accounts for \\SVRPRINC00


-------------------------------------------------------------------------------
*Cloneable Domain Controllers
*DnsUpdateProxy
*Domain Admins
*Domain Computers
*Domain Controllers
*Domain Guests
*Domain Users
*Enterprise Admins
*Enterprise Read-only Domain Controllers
```

```
*Group Policy Creator Owners
*Protected Users
*Read-only Domain Controllers
*Schema Admins
The command completed successfully..
```

## STEP 4: Add a group NET GROUP

a) Add a group to the operating system within a domain, such as Security Group. No permissions are set.

```
C:\Windows\SYSVOL\sysvol\bspWeb.local>NET    GROUP    SMR22      /ADD    /DOMAIN
The command completed successfully.

C:\Windows\SYSVOL\sysvol\bspWeb.local>NET    GROUP    SEGUNDO    /ADD    /DOMAIN
The command completed successfully.

C:\Windows\SYSVOL\sysvol\bspWeb.local>net group

Group Accounts for \\SVRPRINC00

-------------------------------------------------------------------------------
*Cloneable Domain Controllers
*DnsUpdateProxy
*Domain Admins
*Domain Computers
*Domain Controllers
*Domain Guests
*Domain Users
*Enterprise Admins
*Enterprise Read-only Domain Controllers
*Group Policy Creator Owners
*Protected Users
*Read-only Domain Controllers
*Schema Admins
*SEGUNDO
*SMR22
The command completed successfully.
```

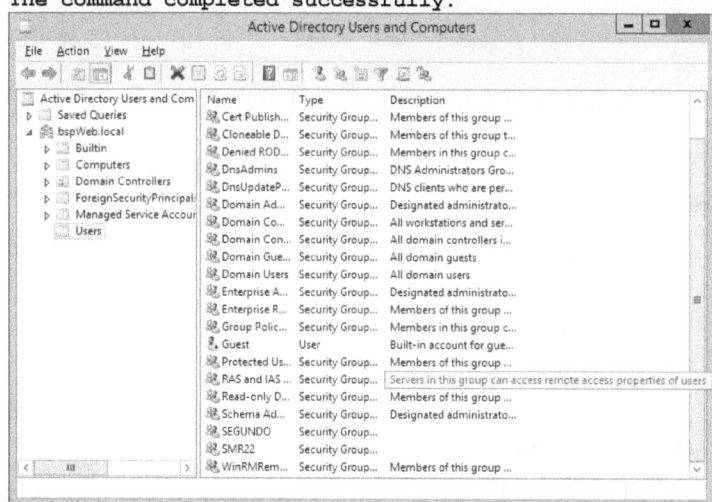

Sever Manager -> Tools
Active Directory Users and Computer

## STEP 5: Add / display / delete local groups NET LOCALGROUP

a) Help in the command line NET LOCALGROUP.
```
C:\Windows\system32>net localgroup  /?
```
b) Display local groups.
```
C:\Windows\SYSVOL\sysvol\bspWeb.local>net localgroup

Aliases for \\SVRPRINC00
-------------------------------------------------
*Access Control Assistance Operators
*Account Operators
*Administrators
*Allowed RODC Password Replication Group
*Backup Operators
*Cert Publishers
*Certificate Service DCOM Access
*Cryptographic Operators
*Denied RODC Password Replication Group
*Distributed COM Users
*DnsAdmins
*Event Log Readers
*Guests
*Hyper-V Administrators
*IIS_IUSRS
```

NET LOCALGROUP [grupo [/COMMENT:"texto"]] [/DOMAIN]
    grupo {/ADD [/COMMENT:"texto"] | /DELETE} [/DOMAIN]
        grupo nombre [...] {/ADD | /DELETE} [/DOMAIN]

```
*Incoming Forest Trust Builders
*Network Configuration Operators
*Performance Log Users
*Performance Monitor Users
*Pre-Windows 2000 Compatible Access
*Print Operators
*RAS and IAS Servers
*RDS Endpoint Servers
*RDS Management Servers
*RDS Remote Access Servers
*Remote Desktop Users
*Remote Management Users
*Replicator
*Server Operators
*Terminal Server License Servers
*Users
*Windows Authorization Access Group
*WinRMRemoteWMIUsers__
The command completed successfully.
```

c) Add a local group plus a comment, the comment must be enclosed in double quotes.
```
C:\Windows\system32>net localgroup  AULOS-APROBADOS   /COMMENT:"MIS ALUMNOS APROBADOS"   /ADD
The command completed successfully.
C:\Windows\system32>net localgroup AULOS-APROBADOS-ASIR /ADD
The command completed successfully.
```

d) Consult local groups, operating system.
```
C:\Windows\system32>NET LOCALGROUP

Alias for \\SVR-BSP-00
-----------------------------------------------------------------------------
* Administrators
* Administrators DHCP
* Hyper-V Administrators
* AULOS-APPROVED
* AULOS-APPROVED-GRAB
* Certificate Service DCOM Access
* Duplicators
* IIS_IUSRS
* Guests
* Event Log Readers
* Cryptographic Operators
* Assistance Operators access control
* Network Configuration Operators
* Backup Operators
* Opers. of impression
* RDS Remote Access Servers
* RDS Management Servers
* RDS End Servers
* Users
* Power Users
* Distributed COM Users
* Remote Management Users
* DHCP users
* Remote Desktop Users
* Users Performance Monitor
* Performance Log Users
* WinRMRemoteWMIUsers__
The command completed successfully.
```

e) Delete a local group. You must specify the full name.
```
C:\Windows\system32>net localgroup AULOS-APROBADOS-ASIR /DELETE
The command completed successfully.
```

f) Check that has deleted the local group, enough to show the local groups that no longer exists.
```
C:\Windows\system32>NET LOCALGROUP

Alias for \\SVR-BSP-00
-----------------------------------------------------------------------------
* Administrators
* Administrators DHCP
* Hyper-V Administrators
* AULOS-APPROVED
* Certificate Service DCOM Access
* Duplicators
* IIS_IUSRS
* Guests
* Event Log Readers
* Cryptographic Operators
* Assistance Operators access control
* Network Configuration Operators
* Backup Operators
* Opers. of impression
* RDS Remote Access Servers
```

```
* RDS Management Servers
* RDS End Servers
* Users
* Power Users
* Distributed COM Users
* Remote Management Users
* DHCP users
* Remote Desktop Users
* Users Performance Monitor
* Performance Log Users
* WinRMRemoteWMIUsers__
The command completed successfully.
```

## STEP 6: Add / view / delete users Domain NET USER

### STEP 6.1: Display Domain Users

```
NET  USER
C:\Windows\SYSVOL\sysvol\bspWeb.local>net user
C:\Windows\SYSVOL\sysvol\bspWeb.local>net user

User accounts for \\SVRPRINC00
-------------------------------------------------------------------------------
Administrator            Guest                    krbtgt
The command completed successfully.
```

### STEP 6.2: Create users in the domain

```
NET    USER    RUBEN      /ADD    /DOMAIN
NET    USER    ADRIAN     /ADD    /DOMAIN
NET    USER    JORGE      /ADD    /DOMAIN
NET    USER    ASER       /ADD    /DOMAIN
```

> **NOTE:** Display the error code: NET HELPMSG código

```
C:\Windows\SYSVOL\sysvol\bspWeb.local>NET    USER    RUBEN    /ADD  /DOMAIN
 The password does not meet the requirements of the password policy. Check the re-
quirements of minimum length, complexity and password history.
You can get more help with command NET HELPMSG 2245.
```

## STEP 7: Display the error code NET HELPMSG

```
NET  HELPMSG  2245
C:\Windows\SYSVOL\sysvol\bspWeb.local>NET    HELPMSG    2245
The password does not meet the requirements of the password policy. Check the requirements
of minimum length, complexity and password history.
```

## STEP 8: Register a user by specifying the key NET USER

```
C:\Windows\SYSVOL\sysvol\bspWeb.local>NET   USER    RUBEN    Practice2016.  /ADD /DOMAIN
The command completed successfully.

C:\Windows\SYSVOL\sysvol\bspWeb.local>NET   USER   ADRIAN    Practice2016. /ADD   /DOMAIN
The command completed successfully.

C:\Windows\SYSVOL\sysvol\bspWeb.local>NET   USER    JORGE    Practice2016. /ADD /DOMAIN
The command completed successfully.

C:\Windows\SYSVOL\sysvol\bspWeb.local>NET   USER    ASER     Practice2016. /ADD    /DOMAIN
The command completed successfully.
```

### STEP 8.1: Create random keys NET USER ... / RANDOM

a) Create random keys while the user is created in the system.

```
NET USER   ANA    /ADD /DOMAIN /RANDOM
C:\Windows\SYSVOL\sysvol\bspWeb.local>NET USER    ANA /RANDOM /ADD /DOMAIN
Password for ANA is: ogMdQM%1

The command completed successfully.
```

b) Must store the keys assigned directly to users to be registered and can be modified.

```
C:\Windows\SYSVOL\sysvol\bspWeb.local>NET USER    ANA /RANDOM /ADD /DOMAIN  >>
CLAVES.TXT
         NET      USER
```

c) Check high data users in the domain.

```
C:\Windows\SYSVOL\sysvol\bspWeb.local>net user

User accounts for \\SVRPRINC00

-------------------------------------------------------------------------------
Administrator            ADRIAN                   ANA
ASER                     Guest                    JORGE
krbtgt                   RUBEN
The command completed successfully.
```

# STEP 9: Delete User, Groups and Resources.
## STEP 9.1: Remove a domain user. NET USER ... /DELETE
a) Deleting a user (assuming they exist).
```
C:\Windows\SYSVOL\sysvol\bspWeb.local>NET   USER   ANA   /DELETE
The command completed successfully.
```
b) Check that the user is deleted.
```
C:\Windows\SYSVOL\sysvol\bspWeb.local>net user
User accounts for \\SVRPRINC00

-------------------------------------------------------------------------------
Administrator              ADRIAN                    ASER
Guest                      JORGE                     krbtgt
RUBEN
The command completed successfully.
```
## STEP 9.2: Deleting a domain group. GROUP NET ... / DELETE
a) Previously the groups are displayed and notes that the group exists to delete.
NET GROUP
```
C:\Windows\SYSVOL\sysvol\bspWeb.local>NET GROUP

Group Accounts for \\SVRPRINC00

-------------------------------------------------------------------------------
*Cloneable Domain Controllers
*DnsUpdateProxy
*Domain Admins
*Domain Computers
*Domain Controllers
*Domain Guests
*Domain Users
*Enterprise Admins
*Enterprise Read-only Domain Controllers
*Group Policy Creator Owners
*Protected Users
*Read-only Domain Controllers
*Schema Admins
*SEGUNDO
*SMR22
The command completed successfully.
```

b) Delete the group we have seen that already exists.
NET GROUP   TEMPORAL  /ADD /DOMAIN
```
C:\Windows\SYSVOL\sysvol\bspWeb.local>net group temporal /add /domain
The command completed successfully.

C:\Windows\SYSVOL\sysvol\bspWeb.local>net group

Group Accounts for \\SVRPRINC00
-------------------------------------------------------------------------------
*Cloneable Domain Controllers
*DnsUpdateProxy
*Domain Admins
*Domain Computers
*Domain Controllers
*Domain Guests
*Domain Users
*Enterprise Admins
*Enterprise Read-only Domain Controllers
*Group Policy Creator Owners
*Protected Users
*Read-only Domain Controllers
*Schema Admins
*SEGUNDO
*SMR22
*temporal
The command completed successfully..
```

b.1) Delete group.
```
C:\Windows\SYSVOL\sysvol\bspWeb.local>net group temporal /delete
The command completed successfully.


C:\Windows\SYSVOL\sysvol\bspWeb.local>net group

Group Accounts for \\SVRPRINC00

-------------------------------------------------------------------------------
*Cloneable Domain Controllers
*DnsUpdateProxy
*Domain Admins
*Domain Computers
```

```
                *Domain Controllers
                *Domain Guests
                *Domain Users
                *Enterprise Admins
                *Enterprise Read-only Domain Controllers
                *Group Policy Creator Owners
                *Protected Users
                *Read-only Domain Controllers
                *Schema Admins
                *SEGUNDO
                *SMR22
                The command completed successfully.
```

c) Display information of all shares.
```
        NET VIEW %LOGONSERVER% /ALL
        C:\Windows\SYSVOL\sysvol\bspWeb.local>NET VIEW   %LOGONSERVER%   /ALL
        Shared resources at \\SVRPRINC00

        Share name       Type  Used as  Comment
        -------------------------------------------------------------------------------
        ADMIN$           Disk           Remote Admin
        C$               Disk           Default share
        CARPETAUSERIOS   Disk
        IPC$             IPC            Remote IPC
        NETLOGON         Disk           Logon server share
        REPASO           Disk  M:
        SYSVOL           Disk           Logon server share
        The command completed successfully.
```

d) Display information shared resources on the server.
```
        C:\Windows\SYSVOL\sysvol\bspWeb.local>net use
        New connections will be remembered.

        Status       Local     Remote                   Network
        -------------------------------------------------------------------------------
        Disconnected M:        \\SVRPRINC00\REPASO      Microsoft Windows Network
        The command completed successfully.
```

e) Display only shared resources.
```
        C:\Windows\SYSVOL\sysvol\bspWeb.local>NET SHARE

        Share name    Resource                              Remark

        -------------------------------------------------------------------------------
        C$            C:\                                   Default share
        IPC$                                                Remote IPC
        ADMIN$        C:\Windows                            Remote Admin
        CARPETAUSERIOS
                      C:\Windows\SYSVOL\domain\REP...
        NETLOGON      C:\Windows\SYSVOL\sysvol\bspWeb.local\SCRIPTS
                                                            Logon server share
        REPASO        C:\Windows\SYSVOL\domain\REP...
        SYSVOL        C:\Windows\SYSVOL\sysvol              Logon server share
        The command completed successfully.
```

## STEP 9.3: Deleting a shared resource . NET USE ... /DELETE
```
        NET USE  M:  /DELETE
        NET USE
        C:\Windows\SYSVOL\sysvol\bspWeb.local>NET USE   M:    /DELETE
        M: was deleted successfully.

        C:\Windows\SYSVOL\sysvol\bspWeb.local>NET USE
        New connections will be remembered.

        There are no entries in the list.
```

## STEP 10: Display domain information.

### STEP 10.1: Display information about the domain NET VIEW.
```
        NET VIEW %LOGONSERVER% /DOMAIN:bspweb.local
        WHOAMI /FQDN
```
a) Display all information that has a cache server connections.
```
        C:\Windows\SYSVOL\sysvol\bspWeb.local>net view  \\svrprinc00  /cache
        Shared resources at \\svrprinc00

        Share name       Type  Used as  Comment
        -------------------------------------------------------------------------------
        CARPETAUSERIOS   Disk           Manual caching of documents
```

```
NETLOGON         Disk            Manual caching of documents
REPASO           Disk            Manual caching of documents
SYSVOL           Disk            Manual caching of documents
The command completed successfully.

C:\Windows\SYSVOL\sysvol\bspWeb.local>net view  \\svrprinc00  /cache  /all
Shared resources at \\svrprinc00

Share name       Type  Used as   Comment
-------------------------------------------------------------------------------
ADMIN$           Disk            Manual caching of documents
C$               Disk            Manual caching of documents
CARPETAUSERIOS   Disk            Manual caching of documents
IPC$             IPC             Manual caching of documents
NETLOGON         Disk            Manual caching of documents
REPASO           Disk            Manual caching of documents
SYSVOL           Disk            Manual caching of documents
The command completed successfully.
```

b) Display all information of a domain server.
```
C:\Windows\SYSVOL\sysvol\bspWeb.local>net view  \\svrprinc00    /all
Shared resources at \\svrprinc00

Share name       Type  Used as   Comment
-------------------------------------------------------------------------------
ADMIN$           Disk            Remote Admin
C$               Disk            Default share
CARPETAUSERIOS   Disk
IPC$             IPC             Remote IPC
NETLOGON         Disk            Logon server share
REPASO           Disk
SYSVOL           Disk            Logon server share
The command completed successfully.
```

c) Display all information of a domain server.
```
C:\Windows\SYSVOL\sysvol\bspWeb.local>net view  \\svrprinc00    /all /domain:bsp
Web.local
Shared resources at \\svrprinc00

Share name       Type  Used as   Comment
-------------------------------------------------------------------------------
ADMIN$           Disk            Remote Admin
C$               Disk            Default share
CARPETAUSERIOS   Disk
IPC$             IPC             Remote IPC
NETLOGON         Disk            Logon server share
REPASO           Disk
SYSVOL           Disk            Logon server share
The command completed successfully.
```

**STEP 10.2: Display information about connections to the domain NET SESSION**

a) List sessions or open sessions.
   NET SESSION
b) Open sessions within the server.
   NET SESSION %LOGONSERVER%
```
C:\Windows\SYSVOL\sysvol\bspWeb.local>net session
There are no entries in the list.

C:\Windows\SYSVOL\sysvol\bspWeb.local>NET SESSION   %LOGONSERVER%
A session does not exist with that computer name.

More help is available by typing NET HELPMSG 2312.
```

c) Display all connections and how long they have open.
   NET SESSION /LIST
```
C:\Windows\SYSVOL\sysvol\bspWeb.local>NET     SESSION     /LIST
Username                    SVRPRINC00 $
Computer Name               [:: 1]
Start as a guest            No
Customer type
Downtime                    00:00:10

Command is completed successfully
```

d) Close the connection of a user from a particular machine.
   NET SESSION \\PUESTO05   /DELETE

## STEP 11: Add posts (computers) to the domain NET COMPUTER

a) Add the ports from the command line.

```
C:\Windows\SYSVOL\sysvol\bspWeb.local>NET    COMPUTER    \\PUESTO05    /ADD
The command completed successfully.

C:\Windows\SYSVOL\sysvol\bspWeb.local>NET    COMPUTER    \\PUESTO04    /ADD
The command completed successfully.

C:\Windows\SYSVOL\sysvol\bspWeb.local>NET    COMPUTER    \\PUESTO06    /ADD
The command completed successfully.
```

b) Display from the graphical environment aggregates jobs from the command line.

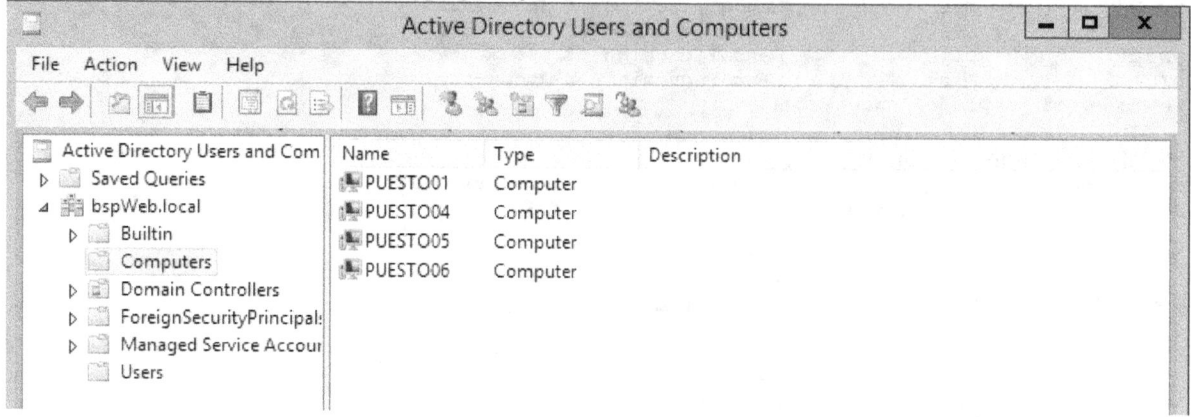

Clear computer from a domain.
NET COMPUTER  \\PUESTO04  /DELETE

```
C:\Windows\SYSVOL\sysvol\bspWeb.local>NET COMPUTER    \\PUESTO04    /DELETE
The command completed successfully.
```

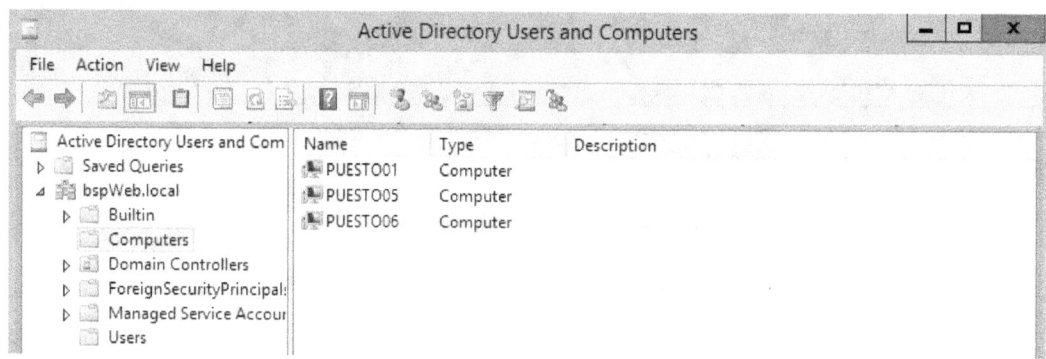

## STEP 12: List all entries connection files. NET FILE

NET  FILE
```
C:\Windows\SYSVOL\sysvol\bspWeb.local>NET FILE
There are no entries in the list.
```

a) Close tickets, according to identification.
NET  FILE  numid  /CLOSE

## STEP 13: Display Server configuration information and Stations

### STEP 13.1: Display information NET CONFIG

a) Display configuration information.
NET  CONFIG
b) Display server configuration.
NET CONFIG  SERVER
```
C:\Windows\SYSVOL\sysvol\bspWeb.local>NET CONFIG SERVER
Server Name                          \\SVRPRINC00
Server Comment

Software version                     Windows Server 2012 R2 Standard Evaluation

Server is active on
        NetbiosSmb (SVRPRINC00)
        NetBT_Tcpip_{0D7F8ED4-1EC1-4BB3-A058-D0C317F404A5} (SVRPRINC00)

Server hidden                        No
```

```
Maximum Logged On Users            16777216
Maximum open files per session     16384

Idle session time (min)            15
The command completed successfully.
```

## STEP 13.2: Display information workstations NET CONFIG WORKSTATION
```
NET CONFIG  WORKSTATION
C:\Windows\SYSVOL\sysvol\bspWeb.local>NET CONFIG WORKSTATION
Computer name                          \\SVRPRINC00
Full Computer name                     SVRPRINC00.bspWeb.local
User name                              Administrator

Workstation active on
        NetBT_Tcpip_{0D7F8ED4-1EC1-4BB3-A058-D0C317F404A5} (00155D016306)

Software version                       Windows Server 2012 R2 Standard Evaluation

Workstation domain                     BSPWEB0
Workstation Domain DNS Name            bspWeb.local
Logon domain                           BSPWEB0

COM Open Timeout (sec)                 0
COM Send Count (byte)                  16
COM Send Timeout (msec)                250
The command completed successfully.
```

## STEP 13.3: Display information services. NET CONTINUE
a) List the services.
```
C:\Windows\system32>sc query
. . . . . . . . . . . . . . . . . .
SERVICE_NAME: WinRM
DISPLAY_NAME: Windows Remote Management (WS-Management)
        TYPE               : 20  WIN32_SHARE_PROCESS
        STATE              : 4   RUNNING
                                 (STOPPABLE, NOT_PAUSABLE, ACCEPTS_SHUTDOWN)
        WIN32_EXIT_CODE    : 0   (0x0)
        SERVICE_EXIT_CODE  : 0   (0x0)
        CHECKPOINT         : 0x0
        WAIT_HINT          : 0x0

SERVICE_NAME: WLMS
DISPLAY_NAME: Windows Licensing Monitoring Service
        TYPE               : 10  WIN32_OWN_PROCESS
        STATE              : 4   RUNNING
                                 (NOT_STOPPABLE, NOT_PAUSABLE, ACCEPTS_SHUTDOWN)
        WIN32_EXIT_CODE    : 0   (0x0)
        SERVICE_EXIT_CODE  : 0   (0x0)
        CHECKPOINT         : 0x0
        WAIT_HINT          : 0x0
```
b) Display the status of a service.
```
C:\Windows\SYSVOL\sysvol\bspWeb.local>NET CONTINUE   WINS

The WINS service was continued successfully.

C:\Windows\SYSVOL\sysvol\bspWeb.local>NET CONTINUE SPOOLER
The requested pause, continue, or stop is not valid for this service.

More help is available by typing NET HELPMSG 2191.
```

## STEP 13.4: Display information services. NET PAUSE
a) Pause service.
```
C:\Windows\system32> NET PAUSE WINS
The service was stopped successfully WINDOWS.
```
b) Resuming a paused service.
```
C:\Windows\system32>NET CONTINUE WINS
WINS service continued successfully.
```

## STEP 13.5: Synchronize the time with the time server. NET TIME
a) To force a computer to synchronize its time with a specific computer.
```
C:\Windows\system32>NET TIME %LOGONSERVER% /SET
The current time in \\ SVR-BSP-00 is 12/06/2016 4:58:14

The current local clock is 6/12/2016 4:58:14
Do you want to set the local time of the machine to match
the time in \\ SVR-BSP-00? (S / N) [S]: S
Is completed correctly command.
```
b) Another way to synchronize.

Page. 39

```
C:\Windows\system32>NET TIME \\192.168.2.40 /SET /Y
The current time in \\ 192.168.2.40 is 12/06/2016 5:01:47

Command is completed successfully.
```

## STEP 13.6: Display network information and network objects. DSQUERY
DSQUERY

a) Display information of all objects.
   DSQUERY  *

b) Obtain information on the domain controller.
   DSQUERY  SITE
   WHOAMI   /FQDN

```
C:\Windows\SYSVOL\sysvol\bspWeb.local>WHOAMI       /FQDN
CN=Administrador,CN=Users,DC=bspWeb,DC=local


C:\Windows\SYSVOL\sysvol\bspWeb.local>DSQUERY  SITE
"CN=Default-First-Site-Name,CN=Sites,CN=Configuration,DC=bspWeb,DC=local"
```

NOTE: does not work, if it has not been promoted to Active Directory, previously.

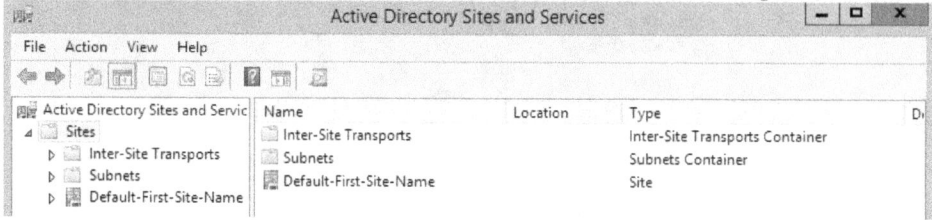

## STEP 13.7: Display contact details domain administrator. DSQUERY CONTACT
DSQUERY CONTACT

## STEP 13.8: Display information about domain computers. DSQUERY COMPUTER
```
C:\Windows\SYSVOL\sysvol\bspWeb.local>DSQUERY COMPUTER
"CN=SVRPRINC00,OU=Domain Controllers,DC=bspWeb,DC=local"
"CN=PUESTO05,CN=Computers,DC=bspWeb,DC=local"
"CN=PUESTO06,CN=Computers,DC=bspWeb,DC=local"
```

## STEP 13.9: Delete the group we have seen that already exists. DSQUERY  GROUP
```
C:\Windows\SYSVOL\sysvol\bspWeb.local>DSQUERY COMPUTER
"CN=SVRPRINC00,OU=Domain Controllers,DC=bspWeb,DC=local"
"CN=PUESTO05,CN=Computers,DC=bspWeb,DC=local"
"CN=PUESTO06,CN=Computers,DC=bspWeb,DC=local"

 C:\Windows\SYSVOL\sysvol\bspWeb.local>DSQUERY GROUP
"CN=WinRMRemoteWMIUsers__,CN=Users,DC=bspWeb,DC=local"
"CN=Administrators,CN=Builtin,DC=bspWeb,DC=local"
"CN=Users,CN=Builtin,DC=bspWeb,DC=local"
"CN=Guests,CN=Builtin,DC=bspWeb,DC=local"
"CN=Print Operators,CN=Builtin,DC=bspWeb,DC=local"
"CN=Backup Operators,CN=Builtin,DC=bspWeb,DC=local"
"CN=Replicator,CN=Builtin,DC=bspWeb,DC=local"
"CN=Remote Desktop Users,CN=Builtin,DC=bspWeb,DC=local"
"CN=Network Configuration Operators,CN=Builtin,DC=bspWeb,DC=local"
"CN=Performance Monitor Users,CN=Builtin,DC=bspWeb,DC=local"
"CN=Performance Log Users,CN=Builtin,DC=bspWeb,DC=local"
"CN=Distributed COM Users,CN=Builtin,DC=bspWeb,DC=local"
"CN=IIS_IUSRS,CN=Builtin,DC=bspWeb,DC=local"
"CN=Cryptographic Operators,CN=Builtin,DC=bspWeb,DC=local"
"CN=Event Log Readers,CN=Builtin,DC=bspWeb,DC=local"
"CN=Certificate Service DCOM Access,CN=Builtin,DC=bspWeb,DC=local"
"CN=RDS Remote Access Servers,CN=Builtin,DC=bspWeb,DC=local"
"CN=RDS Endpoint Servers,CN=Builtin,DC=bspWeb,DC=local"
"CN=RDS Management Servers,CN=Builtin,DC=bspWeb,DC=local"
"CN=Hyper-V Administrators,CN=Builtin,DC=bspWeb,DC=local"
"CN=Access Control Assistance Operators,CN=Builtin,DC=bspWeb,DC=local"
"CN=Remote Management Users,CN=Builtin,DC=bspWeb,DC=local"
"CN=Domain Computers,CN=Users,DC=bspWeb,DC=local"
"CN=Domain Controllers,CN=Users,DC=bspWeb,DC=local"
"CN=Schema Admins,CN=Users,DC=bspWeb,DC=local"
"CN=Enterprise Admins,CN=Users,DC=bspWeb,DC=local"
"CN=Cert Publishers,CN=Users,DC=bspWeb,DC=local"
"CN=Domain Admins,CN=Users,DC=bspWeb,DC=local"
"CN=Domain Users,CN=Users,DC=bspWeb,DC=local"
"CN=Domain Guests,CN=Users,DC=bspWeb,DC=local"
"CN=Group Policy Creator Owners,CN=Users,DC=bspWeb,DC=local"
"CN=RAS and IAS Servers,CN=Users,DC=bspWeb,DC=local"
"CN=Server Operators,CN=Builtin,DC=bspWeb,DC=local"
"CN=Account Operators,CN=Builtin,DC=bspWeb,DC=local"
"CN=Pre-Windows 2000 Compatible Access,CN=Builtin,DC=bspWeb,DC=local"
```

```
"CN=Incoming Forest Trust Builders,CN=Builtin,DC=bspWeb,DC=local"
"CN=Windows Authorization Access Group,CN=Builtin,DC=bspWeb,DC=local"
"CN=Terminal Server License Servers,CN=Builtin,DC=bspWeb,DC=local"
"CN=Allowed RODC Password Replication Group,CN=Users,DC=bspWeb,DC=local"
"CN=Denied RODC Password Replication Group,CN=Users,DC=bspWeb,DC=local"
"CN=Read-only Domain Controllers,CN=Users,DC=bspWeb,DC=local"
"CN=Enterprise Read-only Domain Controllers,CN=Users,DC=bspWeb,DC=local"
"CN=Cloneable Domain Controllers,CN=Users,DC=bspWeb,DC=local"
"CN=Protected Users,CN=Users,DC=bspWeb,DC=local"
"CN=DnsAdmins,CN=Users,DC=bspWeb,DC=local"
"CN=DnsUpdateProxy,CN=Users,DC=bspWeb,DC=local"
"CN=SMR22,CN=Users,DC=bspWeb,DC=local"
"CN=SEGUNDO,CN=Users,DC=bspWeb,DC=local"
"CN=AULOS-APROBADOS,CN=Users,DC=bspWeb,DC=local"
```

## STEP 13.10: Display information containers organizational units. DSQUERY OU

```
C:\Windows\SYSVOL\sysvol\bspWeb.local>DSQUERY OU
"OU=Domain Controllers,DC=bspWeb,DC=local"
```

## STEP 13.11: Display server information domain. DSQUERY SERVER

```
C:\Windows\SYSVOL\sysvol\bspWeb.local>DSQUERY SERVER
"CN=SVRPRINC00,CN=Servers,CN=Default-First-Site-Name,CN=Sites,CN=Configuration,D
C=bspWeb,DC=local"
```

## STEP 13.12: Display information domain users. DSQUERY USER

```
C:\Windows\SYSVOL\sysvol\bspWeb.local>DSQUERY SERVER
"CN=SVRPRINC00,CN=Servers,CN=Default-First-Site-Name,CN=Sites,CN=Configuration,D
C=bspWeb,DC=local"

C:\Windows\SYSVOL\sysvol\bspWeb.local>DSQUERY USER
"CN=Administrator,CN=Users,DC=bspWeb,DC=local"
"CN=Guest,CN=Users,DC=bspWeb,DC=local"
"CN=krbtgt,CN=Users,DC=bspWeb,DC=local"
"CN=RUBEN,CN=Users,DC=bspWeb,DC=local"
"CN=ADRIAN,CN=Users,DC=bspWeb,DC=local"
"CN=JORGE,CN=Users,DC=bspWeb,DC=local"
"CN=ASER,CN=Users,DC=bspWeb,DC=local"
```

## STEP 13.13: Display information by search sites in the Active Directory. DSQUERY SITE

```
C:\Windows\system32> DSQUERY SITE
"CN=Default-First-Site-Name,CN=Sites,CN=Configuration,DC=dawprog0,DC=local"
```

## STEP 13.14: Find objects partition of Active Directory. DSQUERY PARTITION

```
C:\Windows\SYSVOL\sysvol\bspWeb.local>DSQUERY USER
"CN=Administrator,CN=Users,DC=bspWeb,DC=local"
"CN=Guest,CN=Users,DC=bspWeb,DC=local"
"CN=krbtgt,CN=Users,DC=bspWeb,DC=local"
"CN=RUBEN,CN=Users,DC=bspWeb,DC=local"
"CN=ADRIAN,CN=Users,DC=bspWeb,DC=local"
"CN=JORGE,CN=Users,DC=bspWeb,DC=local"
"CN=ASER,CN=Users,DC=bspWeb,DC=local"
```

## STEP 13.15: Display by Output Format. DSQUERY PARTITION

```
C:\Windows\SYSVOL\sysvol\bspWeb.local>DSQUERY PARTITION -O DN
"CN=Configuration,DC=bspWeb,DC=local"
"DC=bspWeb,DC=local"
"CN=Schema,CN=Configuration,DC=bspWeb,DC=local"
"DC=DomainDnsZones,DC=bspWeb,DC=local"
"DC=ForestDnsZones,DC=bspWeb,DC=local"

C:\Windows\SYSVOL\sysvol\bspWeb.local>DSQUERY PARTITION -O RDN
"Enterprise Configuration"
"BSPWEB0"
"Enterprise Schema"
"4d5e930b-7f7c-48ec-a732-b9a11ffdd427"
"47f1f55f-0ee6-46e0-964a-4cbd4857b7c2"
```

## STEP 13.16: Display by partition objects. DSQUERY PARTITION

```
C:\Windows\SYSVOL\sysvol\bspWeb.local>DSQUERY PARTITION -part DAW*

C:\Windows\SYSVOL\sysvol\bspWeb.local>DSQUERY PARTITION -part BS*
"DC=bspWeb,DC=local"
```

## STEP 13.17: Display by Domain. DSQUERY PARTITION

```
C:\Windows\SYSVOL\sysvol\bspWeb.local>DSQUERY PARTITION -S BSPWEB.LOCAL
"CN=Configuration,DC=bspWeb,DC=local"
"DC=bspWeb,DC=local"
"CN=Schema,CN=Configuration,DC=bspWeb,DC=local"
"DC=DomainDnsZones,DC=bspWeb,DC=local"
```

```
            "DC=ForestDnsZones,DC=bspWeb,DC=local"
```

## STEP 13.18: Display by Server. DSQUERY PARTITION

```
C:\Windows\SYSVOL\sysvol\bspWeb.local>DSQUERY PARTITION -S SVRPRINC00
"CN=Configuration,DC=bspWeb,DC=local"
"DC=bspWeb,DC=local"
"CN=Schema,CN=Configuration,DC=bspWeb,DC=local"
"DC=DomainDnsZones,DC=bspWeb,DC=local"
"DC=ForestDnsZones,DC=bspWeb,DC=local"
```

# PRACTICE 3: Create a user, O, groups, users within groups. DSADD

**DESCRIPTION:**

A command-line tool that is built into Windows Server 2012 R2 and earlier versions WINDOWS SERVER. It is available if you have server role Domain Services Active Directory (AD DS) installed.

To use dsadd, dsadd must run the command from a command prompt with elevated privileges (Administrator level).

To open a command prompt with elevated privileges, click Start, click the right mouse button **Command Prompt**, and then click **Run as administrator**.

DSADD

## STEP 1: Create an OU

DSADD   OU   OU=NUEVO,DC=bspweb,DC=local

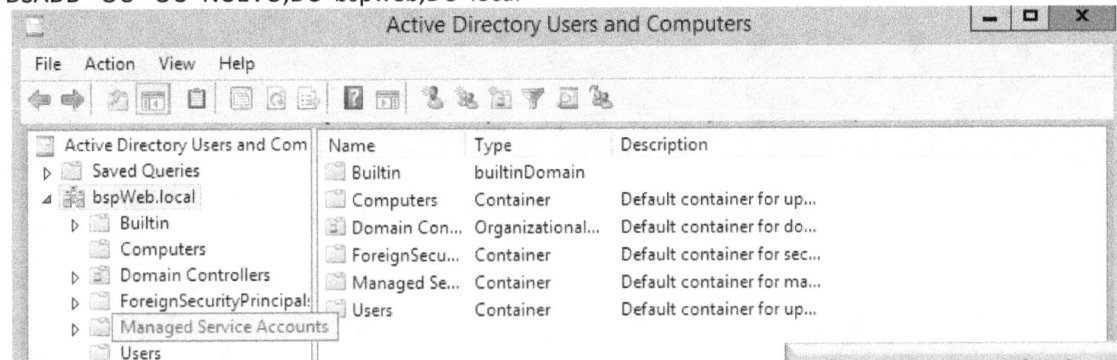

DSADD   OU   OU=PRUEBAS,DC=bspweb,DC=local

**What is an Organizational Unit (OU is abbreviated)?**

It is a container where you can put different Active Directory objects such as Users, Computers, Groups and even other OUs. Among them, we can delegate management permissions on objects we have inside and we can attach domain policies to apply different settings on the types of objects that have inside.(Reference is https://msdn.microsoft.com/es-es/library/jj822947.aspx )

## STEP 2: Create a group within an OU

DSADD   GROUP   CN=TRABAJO,OU=NUEVO,DC=bspweb,DC=local
DSADD   GROUP   CN=TRABAJO2,OU=PRUEBAS,DC=bspweb,DC=local

## STEP 3: Create a user within an organizational unit

DSADD   USER   CN=MOISES,OU=NUEVO,DC=bspweb,DC=local
DSADD   USER   CN=MARCO,OU=NUEVO,DC=bspweb,DC=local
DSADD   USER   CN=DANIEL,OU=PRUEBA,DC=bspweb,DC=local

## STEP 4: Create groups and assign users within a group

DSADD   GROUP   CN=FIESTA,OU=PRUEBA,DC=bspweb,DC=local -members "CN=MOISES   MARCO"

NEW TESTS organizational units are created and then groups are created; WORK and Work2, bassinets, MARCO users are created in the UO = NEW and erroneous evidence of misallocation of a user DANIEL performed.

```
C:\Windows\SYSVOL\sysvol\bspWeb.local>DSADD    OU    OU=NUEVO,DC=bspweb,DC=local
dsadd correcto:OU=NUEVO,DC=bspweb,DC=local

C:\Windows\SYSVOL\sysvol\bspWeb.local>DSADD    OU    OU=PRUEBAS,DC=bspweb,DC=local

dsadd correcto:OU=PRUEBAS,DC=bspweb,DC=local

C:\Windows\SYSVOL\sysvol\bspWeb.local>DSADD    GROUP    CN=TRABAJO,OU=NUEVO,DC=bspw
eb,DC=local
dsadd correcto:CN=TRABAJO,OU=NUEVO,DC=bspweb,DC=local

C:\Windows\SYSVOL\sysvol\bspWeb.local>DSADD    GROUP    CN=TRABAJO2,OU=PRUEBAS,DC=b
spweb,DC=local
dsadd correcto:CN=TRABAJO2,OU=PRUEBAS,DC=bspweb,DC=local

C:\Windows\SYSVOL\sysvol\bspWeb.local>DSADD    USER    CN=MOISES,OU=NUEVO,DC=bspwe
b,DC=local
dsadd correcto:CN=MOISES,OU=NUEVO,DC=bspweb,DC=local

C:\Windows\SYSVOL\sysvol\bspWeb.local>DSADD    USER    CN=MARCO,OU=NUEVO,DC=bspweb
,DC=local
dsadd correcto:CN=MARCO,OU=NUEVO,DC=bspweb,DC=local
```

# PRACTICE 4: Permissions Windows Users.

**DESCRIPTION:**

Each container and network object is assigned access control information. This information is called des-cryptorchid security and controls the type of access allowed to users and groups. Permissions are defined in the security descriptor of an object. They associated or assigned to specific users and groups.

When you are a member of a security group that is associated with an object, you have some ability to manage permissions on that object. For objects that have, control is total. You can use different methods, such as Domain Services Active Directory (AD DS), Group Policy, or access control lists, to manage different types of objects.

## What are permissions

| Special permissions | Total control | Modify | Read and Execute | Display the contents the folder (folders) | Read | Write |
|---|---|---|---|---|---|---|
| Traverse Folder / Execute File | x | x | x | x | | |
| List Folder / Read Data | x | x | x | x | x | |
| Read Attributes | x | x | x | x | x | |
| reading extended attributes | x | x | x | x | x | |
| Create Files / Write Data | x | x | | | | x |
| Create Folders / Append Data | x | x | | | | x |
| Writing attributes | x | x | | | | x |
| Write extended attributes | x | x | | | | x |
| Delete subfolders and files | x | | | | | |
| Remove | x | x | | | | |
| Read Permissions | x | x | x | x | x | x |
| change permissions | x | | | | | |
| Take possession | x | | | | | |
| Sync up | x | x | x | x | x | x |

## Share permissions and NTFS share on a file server

It alpaca: Windows 7, Windows Server 2008 R2 and posteriors.

In a file server, access to a folder can be determined by two sets of permission entries: the share permissions Shared defined in a folder and the NTFS permissions set on the folder (which also can be defined in the archives). The shared resource permissions are often used for managing computers with FAT32 file systems, or other computers that do not use the NTFS file system.

The share permissions and NTFS permissions are independent in the sense that neither changes the other. The final access permissions on a shared folder are determined taking into account inputs permit share and NTFS permission. the most restrictive permissions always apply.

The following table equivalent permissions that an administrator can grant to the Users group for certain shared folder types are proposed. You can also set the share permissions to Full Control for the Everyone group and use NTFS permissions to restrict access.

## Inherited permissions

| Folder Type | Share permissions | NTFS permissions |
|---|---|---|
| Public Folder. A folder that everyone can access. | Grant Change permission to the Users group. | Modify grant permission to the Users group. |
| Private folder. A folder where users can drop confidential reports or homework assignments that can only read the administrator or instructor group. | Grant Change permission to the Users group. | Writing grant permission to the user group that applies to only this folder. If each user needs to have certain permissions for files that lets you can create a permission entry for known security identifier (SID) Creator Owner and apply to subfolders and files only. For example, you can grant Read and Write permissions to the Creator Owner SID in the private folder and apply to all subfolders and files. Thus, the user who has made or created the file (Creator Owner) has the ability to read and write to the file. Then, Creator Owner can access the file using the Run with command:<br>\\ServerName\DropFolder\FileName.<br>Grant Full Control permission to the group manager. |
| Applications folder. A folder containing applications that can be run across the network. | Grant Full Control permission to the group manager. | Grant permissions Read, Read & Execute, and List Folder Contents Users group. |
| Home Folder. individual folder for each user. Only the user has access to the folder. | Read grant permission to the Users group. | Grant Full Control permission to each user on their respective folder. |

## STEP 1: Permissions on objects and resources. ICACLS, CACLS
    ICACLS
    CACLS
    DSACLS

a) Display permissions of all objects.
```
icacls  *.*
UIRibbon.dll NT SERVICE\TrustedInstaller:(F)
             BUILTIN\Administradores:(RX)
             NT AUTHORITY\SYSTEM:(RX)
             BUILTIN\Usuarios:(RX)
             ENTIDAD DE PAQUETES DE APLICACIONES\TODOS LOS PAQUETES DE APLICACIONES:(RX)
...
```

b) Store in a file all the information resulting display the permissions of all objects.
```
ICACLS  *   > RESULTADO.TXT
NOTEPAD  RESULTADO.TXT
```

c) Only search all files that start with AUDIO *
```
C:\Windows\SYSVOL\sysvol\bspWeb.local>ICACLS  R*
REPASO NT AUTHORITY\Authenticated Users:(I)(RX)
       NT AUTHORITY\Authenticated Users:(I)(OI)(CI)(IO)(GR,GE)
       BUILTIN\Server Operators:(I)(RX)
       BUILTIN\Server Operators:(I)(OI)(CI)(IO)(GR,GE)
       BUILTIN\Administrators:(I)(F)
       BUILTIN\Administrators:(I)(OI)(CI)(IO)(F)
       NT AUTHORITY\SYSTEM:(I)(F)
       NT AUTHORITY\SYSTEM:(I)(OI)(CI)(IO)(F)
       CREATOR OWNER:(I)(OI)(CI)(IO)(F)

Successfully processed 1 files; Failed processing 0 files
```

d) Use filters to extract all the .DLL files containing WMS in the name..
```
C:\Windows\system32>icacls   *  |FIND "(F)"|FIND ".DLL"| FIND "WMS"
WMSPDMOD.DLL NT SERVICE\TrustedInstaller:(F)
WMSPDMOE.DLL NT SERVICE\TrustedInstaller:(F)
```

e) Make a copy of all ACLs for all files in c:\windows\system32 and its subdirectories in salidaACL.
```
C:\Windows\system32>icacls   c:\windows\system32\*   /save salidaACL   /T
. . .
c:\windows\system32\config\systemprofile\AppData\Local\Microsoft\Windows\INetCache\Content.IE5\*: Access denied.
7695 files are properly processed; error processing files 1

NOTEPAD   salidaACL
```

```
salidaACL: Bloc de notas
Archivo  Edición  Formato  Ver  Ayuda
0409
D:PAI(A;;FA;;;S-1-5-80-956008885-3418522649-1831038044-1853292631-2271478464)(A;CIIO;GA;;;S-1-5-80-956008885-3418522649-1831038044-1853292631-2271478464)
0C0A
D:PAI(A;;FA;;;S-1-5-80-956008885-3418522649-1831038044-1853292631-2271478464)(A;CIIO;GA;;;S-1-5-80-956008885-3418522649-1831038044-1853292631-2271478464)
7B296FB0-376B-497e-B012-9C450E1B7327-5P-0.C7483456-A289-439d-8115-601632D005A0
D:AI(A;ID;FA;;;SY)(A;ID;FA;;;BA)(A;ID;0x1200a9;;;BU)(A;ID;0x1200a9;;;AC)
7B296FB0-376B-497e-B012-9C450E1B7327-5P-1.C7483456-A289-439d-8115-601632D005A0
D:AI(A;ID;FA;;;SY)(A;ID;FA;;;BA)(A;ID;0x1200a9;;;BU)(A;ID;0x1200a9;;;AC)
@edptoastimage.png
D:PAI(A;;FA;;;S-1-5-80-956008885-3418522649-1831038044-1853292631-2271478464)(A;;0x1200a9;;;BA)(A;;0x1200a9;;;SY)(A;;0x1200a9;;;BU)(A;;0x1200a9;;;AC)S:AI
@language_notification_icon.png
```

f) Check the file created with ACLs.
```
C:\Windows\system32>icacls salidaacl
salidaacl NT AUTHORITY\SYSTEM:(I)(F)
          BUILTIN\Administradores:(I)(F)
          BUILTIN\Usuarios:(I)(RX)
          ENTIDAD DE PAQUETES DE APLICACIONES\TODOS LOS PAQUETES DE APLICACIONES:(I)(RX)

1 records processed correctly; error processing 0 files
```

g) Change the permissions: Grant the user permissions and administrator to delete the file write DAC.
```
C:\Windows\system32>ICACLS   salidaACL   /grant aprendiz:(D,WDAC)
processed file: salidaACL
1 records processed correctly; error processing 0 files

C:\Windows\system32>ICACLS   salidaACL
salidaACL i7-PC\aprendiz:(D,WDAC)
          NT AUTHORITY\SYSTEM:(I)(F)
          BUILTIN\Administradores:(I)(F)
          BUILTIN\Usuarios:(I)(RX)
          ENTIDAD DE PAQUETES DE APLICACIONES\TODOS LOS PAQUETES DE APLICACIONES:(I)(RX)

1 records processed correctly; error processing 0 files
```

## PRACTICE 5: Display AD DS domain information
**DESCRIPTION:**
There are two commands from the command line displays information from Windows Active Directory: dsget, DSQUERY.

**Dsget:** command-line tool that displays information about select properties of a specific object in Active Directory. DSQUERY | DSGET

## STEP 1: DSGET

a) Display users who are members of the Administrators and Users group within the domain bspWeb.local
```
C:\Windows\system32>DSGET USER "CN=Administrador,CN=users,DC=bspWeb,DC=local" -memberof
"CN=Propietarios del creador de directivas de grupo,CN=Users,DC=bspWeb,DC=local"

"CN=Admins. del dominio,CN=Users,DC=bspWeb,DC=local"
"CN=Administradores de empresas,CN=Users,DC=bspWeb,DC=local"
"CN=Administradores de esquema,CN=Users,DC=bspWeb,DC=local"
"CN=Administradores,CN=Builtin,DC=bspWeb,DC=local"
"CN=Usuarios del dominio,CN=Users,DC=bspWeb,DC=local"
```

b) It shows whether the user can change their password or not. Sample: yes or no.
```
C:\Windows\system32>DSGET USER "CN=Administrador,CN=users,DC=bspWeb,DC=local" -canchpwd
  canchpwd
  yes
dsget correcto
```

c) Shows whether the user can change their password or not. Sample: yes or no. Displays the distinguished name (DN) User.
```
C:\Windows\system32>DSGET USER "CN=Administrador,CN=users,DC=bspWeb,DC=local" -canchpwd -dn
  dn                                           canchpwd
  CN=Administrador,CN=users,DC=bspWeb,DC=local  yes
dsget correcto
```

d) Displays whether the user can change their password or not. Sample: yes or no. Displays the id. User Safety.
```
C:\Windows\system32>DSGET USER "CN=Administrador,CN=users,DC=bspWeb,DC=local" -canchpwd -sid
  sid                                              canchpwd
  S-1-5-21-2298800814-2528216055-1890261488-500    yes
dsget correcto
```

e) Shows whether the user can change their password or not. Displays the SAM account name of the user.
```
C:\Windows\system32>DSGET USER "CN=Administrador,CN=users,DC=bspWeb,DC=local" -canchpwd -samid
  samid              canchpwd
  Administrador      yes
dsget correcto
```

f) Displays whether the user can change their password or not. It displays the user principal name.
```
C:\Windows\system32>DSGET USER "CN=Administrador,CN=users,DC=bspWeb,DC=local" -canchpwd -upn
  upn        canchpwd
             yes
dsget correcto
```

g) Displays whether the user can change their password or not. Displays the user name.
```
C:\Windows\system32>DSGET USER "CN=Administrador,CN=users,DC=bspWeb,DC=local" -canchpwd -fn
  fn      canchpwd
          yes
dsget correcto
```

h) Displays whether the user can change their password or not. The second shows the initial user name.
```
C:\Windows\system32>DSGET USER "CN=Administrador,CN=users,DC=bspWeb,DC=local" -canchpwd -mi
  mi      canchpwd
          yes
dsget correcto
```

i) Shows whether the user can change their password or not. Displays user names.
```
C:\Windows\system32>DSGET USER "CN=Administrador,CN=users,DC=bspWeb,DC=local" -canchpwd -ln
  ln      canchpwd
          yes
dsget correcto
```

j) Continuous operation mode: reports errors but continues with the next object in the argument list how-do multiple target objects are specified. Without this option, the command will end on the first error.
```
C:\Windows\system32>DSGET USER "CN=Administrador,CN=users,DC=bspWeb,DC=local" -c
anchpwd -display
  display      canchpwd
               yes
dsget correcto
```

k) Displays whether the user can change their password or not. Displays the id. user's employee.
```
C:\Windows\system32>DSGET USER "CN=Administrador,CN=users,DC=bspWeb,DC=local" -canchpwd -empid
```

```
    empid     canchpwd
              yes
dsget correcto
```

l) Displays the user description.
```
C:\Windows\system32>DSGET USER "CN=Administrador,CN=users,DC=bspWeb,DC=local" -desc
  desc
  Integrated account management for computer or domain
dsget correct
```

m) Displays the user's departmen.
```
C:\Windows\system32>DSGET USER "CN=Administrador,CN=users,DC=bspWeb,DC=local" -dept
  dept

dsget correct
```

n) Displays the user's department and office location User.
```
C:\Windows\system32>DSGET USER "CN=Administrador,CN=users,DC=bspWeb,DC=local" -dept -office
  office    dept

dsget correct
```

o) Displays the department, phone number and email address of the user.
```
C:\Windows\system32>DSGET USER "CN=Administrador,CN=users,DC=bspWeb,DC=local" -dept -tel  -email
   tel      email     dept

dsget correct
```

p) Displays the IP phone number, the fax number of the user and the URL of the website user.
```
C:\Windows\system32>DSGET USER "CN=Administrador,CN=users,DC=bspWeb,DC=local" -iptel  -fax      -webpg
   fax      iptel    webpg

dsget correct
```

q) Display the groups to which the user belongs.
```
C:\Windows\system32>dsget user CN=USER1,CN=users,DC=bspWeb,DC=local   -memberof
"CN=Usuarios del dominio,CN=Users,DC=bspWeb,DC=local"
```

r) Displays the person responsible for the user.
```
C:\Windows\system32>DSGET USER CN=USER1,CN=users,DC=bspWeb,DC=local   -mgr
  mgr
```

s) Displays whether the user account is disabled for logon or not. Sample: yes or no.
```
C:\Windows\system32>dsget user CN=USER1,CN=users,DC=bspWeb,DC=local   -disabled
  disabled
  no
dsget correct
```

t) Displays the directory, the drive letter is assigned to the user's home directory (if the path of the home directory is a UNC path). Shows the particular drive letter of the user (if the parti-cular directory is a UNC path). Displays the path of the user profile.
```
C:\Windows\system32>DSGET USER CN=USER1,CN=users,DC=bspWeb,DC=local   -hmdir  -hmdrv   -profile
   hmdir      hmdrv     profile

dsget correct
```

u) Displays the directory, the drive letter is assigned to the user's home directory (if the path of the home directory is a UNC path). Shows the particular drive letter of the user (if the parti-cular directory is a UNC path). Displays the path of the user profile. Displays the script path logon user.
```
C:\Windows\system32>DSGET USER CN=USER1,CN=users,DC=bspWeb,DC=local   -hmdir  -hmdrv   -profile -loscr
   hmdir      hmdrv     profile    loscr

dsget correct
```

v) Displays the directory, the drive letter is assigned to the user's home directory (if the path of the home directory is a UNC path). Displays whether the user must change their password the next time you log on. Sample: yes or no. Displays whether the user can change their password or not.
```
C:\Windows\system32>DSGET USER CN=USER1,CN=users,DC=bspWeb,DC=local   -hmdir  -mustchpwd -canchpwd
   hmdir    mustchpwd    canchpwd
              no          yes
dsget correct
```

w) Displays the directory, the drive letter is assigned to the user's home directory (if the path of the home directory is a UNC path). Displays whether the user must change their password the next time you log on. Sample: yes or no. And it shows whether the user can change their password or not. Displays if the user password never expires. Sample: yes or no. Displays whether the user account is disabled for logon or not. Sample: yes or no. It shows when the user account expires. Values shown: the date on which the account will expire or chain "never" if the account never expires.

```
C:\Windows\system32>DSGET USER CN=USER1,CN=users,DC=bspWeb,DC=local  -hmdir  -mustchpwd -canchpwd -pwdneverexpires
   hmdir      mustchpwd      canchpwd      pwdneverexpires
              no             yes           no
dsget correct
```

x) Displays the directory, the drive letter is assigned to the user's home directory (if the path of the home directory is a UNC path). Displays whether the user must change their password the next time you log on. Sample: yes or no. Displays whether the user can change their password or not. Displays if the user password never expires. Sample: yes or no. Displays whether the user account is disabled for logon or not. Sample: yes or no.Muestra when the user account expires. Values shown: the date on which the account will expire or chain "never" if the account never expires. account (disable) is deactivated

```
C:\Windows\system32>DSGET USER CN=USER1,CN=users,DC=bspWeb,DC=local  -hmdir  -mustchpwd -canchpwd -pwdneverexpires -disabled
   hmdir      mustchpwd      canchpwd      pwdneverexpires      disabled
              no             yes           no                   no
dsget correct
```

y) Displays the directory, the drive letter is assigned to the user's home directory (if the path of the home directory is a UNC path). Displays whether the user must change their password the next time you log on. Sample: yes or no. Displays whether the user can change their password or not. Displays if the user password never expires. Sample: yes or no. Displays whether the user account is disabled for logon or not. Sample: yes or no. It shows when the user account expires. Values shown: the date on which the account will expire or chain "never" if the account never expires.

```
C:\Windows\system32>DSGET USER CN=USER1,CN=users,DC=bspWeb,DC=local  -hmdir  -mustchpwd -canchpwd -pwdneverexpires -disabled -acctexpires
   hmdir      mustchpwd      canchpwd      pwdneverexpires      acctexpires      disabled

              no             yes           no                   never            no
dsget correct
```

z) Displays the directory, the drive letter is assigned to the user's home directory (if the path of the home directory is a UNC path). Displays whether the user must change their password the next time you log on. Sample: yes or no. Displays whether the user can change their password or not. Displays if the user password never expires. Sample: yes or no. Displays whether the user account is disabled for logon or not. Sample: yes or no.Muestra when the user account expires. Values shown: the date on which the account will expire or chain "never" if the account never expires. Specifies that input from or output channeling towards it is formatted Unicode.

```
C:\Windows\system32>DSGET USER CN=USER1,CN=users,DC=bspWeb,DC=local  -hmdir  -mustchpwd -canchpwd -pwdneverexpires -disabled -acctexpires -uc
   hmdir      mustchpwd      canchpwd      pwdneverexpires      acctexpires      disabled

              no             yes           no                   never            no
dsget correct
```

aa) Displays the directory, the drive letter is assigned to the user's home directory (if the path of the home directory is a UNC path). Displays whether the user must change their password the next time you log on. Sample: yes or no. Displays whether the user can change their password or not. Displays if the user password never expires. Sample: yes or no. Displays whether the user account is disabled for logon or not. Sample: yes or no.Muestra when the user account expires. Values shown: the date on which the account will expire or chain "never" if the account never expires. Specifies that the exit pipe or file is formatted Unicode.

```
C:\Windows\system32>DSGET USER CN=USER1,CN=users,DC=bspWeb,DC=local  -hmdir  -mustchpwd -canchpwd -pwdneverexpires -disabled -acctexpires -uco
?  hmdir      mustchpwd      canchpwd      pwdneverexpires      acctexpires      disabled

              no             yes           no                   never            no
dsget correct
```

bb) Displays the directory, the drive letter is assigned to the user's home directory (if the path of the home directory is a UNC path). Displays whether the user must change their password the next time you log on. Sample: yes or no. Displays whether the user can change their password or not. Displays if the user password never expires. Sample: yes or no. Displays whether the user account is disabled for logon or not. Sample: yes or no.Muestra when the user account expires. Values shown: the date on which the account will expire or chain "never" if the account never expires. It specifies that input from the pipe or file is formatted Unicode.

```
C:\Windows\system32>DSGET USER CN=USER1,CN=users,DC=bspWeb,DC=local  -hmdir  -mustchpwd -canchpwd -pwdneverexpires -disabled -acctexpires -uci
   hmdir      mustchpwd      canchpwd      pwdneverexpires      acctexpires      disabled

              no             yes           no                   never            no
dsget correct
```

cc) ) Displays the directory, the drive letter is assigned to the user's home directory (if the path of the home directory is a UNC path). Displays whether the user must change their password the next time you log on. Sample: yes or no. Displays whether the user can change their password or not. Displays if the user password never expires. Sample: yes or no. Displays whether the user account is disabled for logon or not. Sample: yes or no.Muestra when the user account expires. Values shown: the date on which the account will expire or chain "never" if the account never expires. It connects to the directory partition with the distinguished name <DNDePartición>.
```
C:\Windows\system32>DSGET USER CN=USER1,CN=users,DC=bspWeb,DC=local  -hmdir  -mustchpwd -canchpwd -
pwdneverexpires -disabled -acctexpires -part
dsget failed:No value specified for 'part'.
type dsget /? for help.
```

dd) Displays the directory, the drive letter is assigned to the user's home directory (if the path of the home directory is a UNC path). Displays whether the user must change their password the next time you log on. Sample: yes or no. Displays whether the user can change their password or not. Displays if the user password never expires. Sample: yes or no. Displays whether the user account is disabled for logon or not. Sample: yes or no.Muestra when the user account expires. Values shown: the date on which the account will expire or chain "never" if the account never expires. Displays the effective quota of the user in the specified directory partition.
```
C:\Windows\system32>DSGET USER CN=USER1,CN=users,DC=bspWeb,DC=local  -hmdir  -mustchpwd -canchpwd -
pwdneverexpires -disabled -acctexpires -qlimit
dsget incorrect: The parameter is not correcto.: The directory partition does not exist on the server or
domain specified. Verify That You Entered the correct partition name
Write dsget /? for help.    hmdir     mustchpwd      canchpwd       pwdnever
expires      acctexpires     disabled      qlimit
                no           yes             no                      never            no

dsget correct
```

ee) Displays the directory, the drive letter is assigned to the user's home directory (if the path of the home directory is a UNC path). Displays whether the user must change their password the next time you log on. Sample: yes or no. Displays whether the user can change their password or not. Displays if the user password never expires. Sample: yes or no. Displays whether the user account is disabled for logon or not. Sample: yes or no. It shows when the user account expires. Values shown: the date on which the account will expire or chain "never" if the account never expires. Displays the effective quota of the user in the specified directory partition. Shows which part of the quota used the user in the specified directory partition.
```
C:\Windows\system32>DSGET USER CN=USER1,CN=users,DC=bspWeb,DC=local  -hmdir  -mustchpwd -canchpwd -
pwdneverexpires -disabled -acctexpires -qlimit -qused
dsget incorrect: The parameter is not correcto.: The directory partition does not exist on the server or
domain specified. Verify that you entered the correct partition name
Write dsget /? for incorrect ayuda.dsget: The parameter is not correcto.: The directory partition does
not exist on the server or domain specified. Verify that you entered the correct partition name
Write dsget /? for help.    hmdir     mustchpwd     canchpwd     pwdneverexpires     acctexpires     disabled
qlimit    qused
                no           yes            no                    never                 no

dsget correct
```

## STEP 2: NETDOM
It allows administrators to manage Active Directory domains and trust relationships from the command prompt. Must exist between two different domain trees forming a forest

### STEP 2.1: Check information workstations. NETDOM QUERY WORKSTATION
```
C:\Windows\system32>netdom query workstation
List of workstations in the domain accounts:

PUESTO05 (workstation or server)

PUESTO06 (workstation or server)

The command completed successfully.
```

### STEP 2.2: Check the domain server information. NETDOM QUERY SERVER
```
C:\Windows\system32>NETDOM QUERY SERVER
List of workstations in the domain accounts:

PUESTO05 (workstation or server)

PUESTO06 (workstation or server)

The command completed successfully.
```

### STEP 2.3: Check the list of computer names domain controller . NETDOM QUERY DC
```
C:\Windows\system32>NETDOM QUERY DC
List of domain controllers in the domain accounts:
```

```
SVRPRINC00
SVRPRINC01
The command completed successfully.
```

## STEP 2.4: Consult the list of domain organizational units. NETDOM QUERY OU

```
C:\Windows\system32>NETDOM QUERY OU
The requested API does not work on the remote server.

The command could not be completed correctly.
```

## STEP 2.4: Consult the current list of domain owners FSMO. NETDOM QUERY FSMO

```
C:\Windows\SYSVOL\sysvol\bspWeb.local>NETDOM QUERY FSMO
Schema master              SVRPRINC00.bspWeb.local
Domain naming master       SVRPRINC00.bspWeb.local
PDC                        SVRPRINC00.bspWeb.local
RID pool manager           SVRPRINC00.bspWeb.local
Infrastructure master      SVRPRINC00.bspWeb.local
The command completed successfully.
```

## STEP 2.5: Consult the list of domain trusts. NETDOM QUERY TRUST

```
C:\Windows\SYSVOL\sysvol\bspWeb.local>NETDOM QUERY TRUST
Direction  Trusted\Trusting domain                      Trust type
=========  =======================                      ==========

The command completed successfully.
```

a) Primary Domain Controller domain.

```
C:\Windows\SYSVOL\sysvol\bspWeb.local>NETDOM QUERY PDC
Primary domain controller for the domain:

SVRPRINC00
The command completed successfully.
```

b) List of domain controllers in the domain accounts.
```
C:\Windows\SYSVOL\sysvol\bspWeb.local>NETDOM QUERY DC
List of domain controllers with accounts in the domain:

SVRPRINC00
The command completed successfully.
```

c) List since connect to servers in the domain accounts

```
C:\Windows\SYSVOL\sysvol\bspWeb.local>NETDOM QUERY OU
The requested API is not supported on the remote server.

The command failed to complete successfully.
```

d) List of workstations in the domain accounts.

```
C:\Windows\SYSVOL\sysvol\bspWeb.local>NETDOM QUERY SERVER
List of servers with accounts in the domain:

PUESTO01        ( Workstation or Server )

PUESTO05        ( Workstation or Server )

PUESTO06        ( Workstation or Server )

The command completed successfully.
```

e) Change the domain name server.
```
C:\Windows\SYSVOL\sysvol\bspWeb.local>NETDOM RENAMECOMPUTER SVRPRINC00 /NewName
SVR2017BSP
This operation will rename the computer SVRPRINC00
to SVR2017BSP.

Certain services, such as the Certificate Authority, rely on a fixed machine
name. If any services of this type are running on SVRPRINC00,
then a computer name change would have an adverse impact.

Do you want to proceed (Y or N)?
Y
The computer needs to be restarted in order to complete the operation.

The command completed successfully.
```

> The computer needs to be restarted in order to complete the operation.

f) Check the server name (HOST).
```
C:\Windows\SYSVOL\sysvol\bspWeb.local>HOSTNAME
SVRPRINC00
```

## STEP 3: Join queries using filters: DSQUERY | DSGET

```
C:\Windows\system32>DSQUERY USER -name a* | DSGET USER -dn -desc
  dn                                               desc

  CN=Administrador,CN=Users,DC=bspWeb,DC=local    Integrated account management for computer or do-
main
  CN=ana,CN=Users,DC=bspWeb,DC=local

  CN=ALUMNOASIR,OU=NUEVO,DC=bspWeb,DC=local

  CN=ALUMNODAW,OU=NUEVO,DC=bspWeb,DC=local

  CN=ALUMNOJUAN,OU=NUEVO,DC=bspWeb,DC=local

dsget correcto
```

a) Even using dsquery and dsget together we can get to get the email addresses of users starting for example by **sali***.
DSQUERY USER –samid sali* | DSGET USER -email

# PRACTICE 6: Display object information and network drives domain.

**DESCRIPTION:**

**Dsquery:** command-line tool that enables searches to LDAP as a valid criterion.

To use dsquery, you must run the dsquery command from a command prompt with elevated privileges. To open a command prompt with elevated privileges, the command prompt must be run as Run as administrator.

## STEP 1: Display network information and network objects. DSQUERY

    DSQUERY

a) Display information of all objects.

    DSQUERY  *

b) Obtain information on the domain controller.

    DSQUERY  SITE
    WHOAMI  /FQDN

> NOTE: does not work if it has not been promoted to Active Directory

```
C:\Windows\SYSVOL\sysvol\bspWeb.local> DSQUERY /FQDN
CN=Administrador,CN=Users,DC=bspWeb,DC=local


C:\Windows\SYSVOL\sysvol\bspWeb.local>DSQUERY  SITE
"CN=Default-First-Site-Name,CN=Sites,CN=Configuration,DC=bspWeb,DC=local"
```

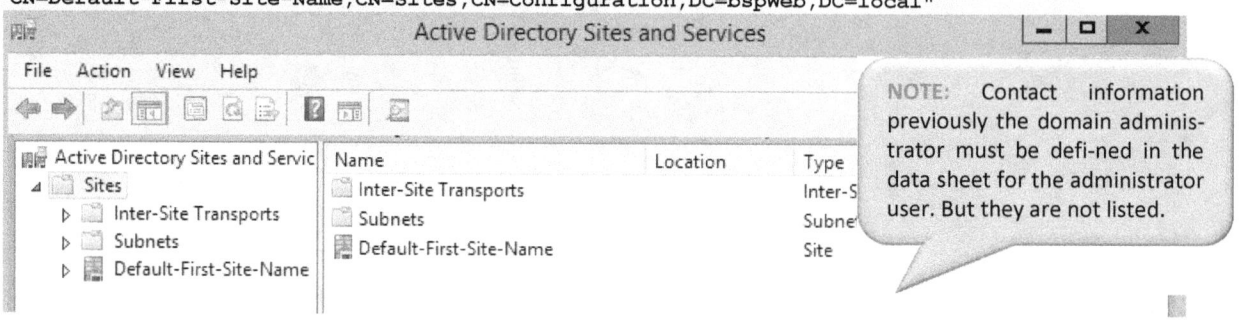

> NOTE: Contact information previously the domain administrator must be defi-ned in the data sheet for the administrator user. But they are not listed.

### STEP 1.1: Display information contact AD DS administrator. DSQUERY CONTACT

    DSQUERY CONTACT

### STEP 1.2: Display information about domain computers. DSQUERY COMPU

    DSQUERY  COMPUTER

```
C:\Windows\SYSVOL\sysvol\bspWeb.local>DSQUERY COMPUTER
"CN=SVRPRINC00,OU=Domain Controllers,DC=bspWeb,DC=local"
"CN=PUESTO05,CN=Computers,DC=bspWeb,DC=local"
"CN=PUESTO06,CN=Computers,DC=bspWeb,DC=local"
n.5) Visualizar información sobre los grupos
        DSQUERY     GROUP


C:\Windows\SYSVOL\sysvol\bspWeb.local>DSQUERY COMPUTER
"CN=SVRPRINC00,OU=Domain Controllers,DC=bspWeb,DC=local"
"CN=PUESTO05,CN=Computers,DC=bspWeb,DC=local"
"CN=PUESTO06,CN=Computers,DC=bspWeb,DC=local"


C:\Windows\SYSVOL\sysvol\bspWeb.local>dsquery contact

C:\Windows\SYSVOL\sysvol\bspWeb.local>DSQUERY GROUP
"CN=WinRMRemoteWMIUsers__,CN=Users,DC=bspWeb,DC=local"
"CN=Administrators,CN=Builtin,DC=bspWeb,DC=local"
"CN=Users,CN=Builtin,DC=bspWeb,DC=local"
"CN=Guests,CN=Builtin,DC=bspWeb,DC=local"
"CN=Print Operators,CN=Builtin,DC=bspWeb,DC=local"
"CN=Backup Operators,CN=Builtin,DC=bspWeb,DC=local"
"CN=Replicator,CN=Builtin,DC=bspWeb,DC=local"
"CN=Remote Desktop Users,CN=Builtin,DC=bspWeb,DC=local"
"CN=Network Configuration Operators,CN=Builtin,DC=bspWeb,DC=local"
"CN=Performance Monitor Users,CN=Builtin,DC=bspWeb,DC=local"
"CN=Performance Log Users,CN=Builtin,DC=bspWeb,DC=local"
"CN=Distributed COM Users,CN=Builtin,DC=bspWeb,DC=local"
"CN=IIS_IUSRS,CN=Builtin,DC=bspWeb,DC=local"
"CN=Cryptographic Operators,CN=Builtin,DC=bspWeb,DC=local"
"CN=Event Log Readers,CN=Builtin,DC=bspWeb,DC=local"
"CN=Certificate Service DCOM Access,CN=Builtin,DC=bspWeb,DC=local"
"CN=RDS Remote Access Servers,CN=Builtin,DC=bspWeb,DC=local"
"CN=RDS Endpoint Servers,CN=Builtin,DC=bspWeb,DC=local"
"CN=RDS Management Servers,CN=Builtin,DC=bspWeb,DC=local"
"CN=Hyper-V Administrators,CN=Builtin,DC=bspWeb,DC=local"
"CN=Access Control Assistance Operators,CN=Builtin,DC=bspWeb,DC=local"
"CN=Remote Management Users,CN=Builtin,DC=bspWeb,DC=local"
"CN=Domain Computers,CN=Users,DC=bspWeb,DC=local"
"CN=Domain Controllers,CN=Users,DC=bspWeb,DC=local"
"CN=Schema Admins,CN=Users,DC=bspWeb,DC=local"
```

> NOTE: If it is a midrange server, the computer BIOS own contact details are previously defined with managed-dor.

```
"CN=Enterprise Admins,CN=Users,DC=bspWeb,DC=local"
"CN=Cert Publishers,CN=Users,DC=bspWeb,DC=local"
"CN=Domain Admins,CN=Users,DC=bspWeb,DC=local"
"CN=Domain Users,CN=Users,DC=bspWeb,DC=local"
"CN=Domain Guests,CN=Users,DC=bspWeb,DC=local"
"CN=Group Policy Creator Owners,CN=Users,DC=bspWeb,DC=local"
"CN=RAS and IAS Servers,CN=Users,DC=bspWeb,DC=local"
"CN=Server Operators,CN=Builtin,DC=bspWeb,DC=local"
"CN=Account Operators,CN=Builtin,DC=bspWeb,DC=local"
"CN=Pre-Windows 2000 Compatible Access,CN=Builtin,DC=bspWeb,DC=local"
"CN=Incoming Forest Trust Builders,CN=Builtin,DC=bspWeb,DC=local"
"CN=Windows Authorization Access Group,CN=Builtin,DC=bspWeb,DC=local"
"CN=Terminal Server License Servers,CN=Builtin,DC=bspWeb,DC=local"
"CN=Allowed RODC Password Replication Group,CN=Users,DC=bspWeb,DC=local"
"CN=Denied RODC Password Replication Group,CN=Users,DC=bspWeb,DC=local"
"CN=Read-only Domain Controllers,CN=Users,DC=bspWeb,DC=local"
"CN=Enterprise Read-only Domain Controllers,CN=Users,DC=bspWeb,DC=local"
"CN=Cloneable Domain Controllers,CN=Users,DC=bspWeb,DC=local"
"CN=Protected Users,CN=Users,DC=bspWeb,DC=local"
"CN=DnsAdmins,CN=Users,DC=bspWeb,DC=local"
"CN=DnsUpdateProxy,CN=Users,DC=bspWeb,DC=local"
"CN=SMR22,CN=Users,DC=bspWeb,DC=local"
"CN=SEGUNDO,CN=Users,DC=bspWeb,DC=local"
"CN=AULOS-APROBADOS,CN=Users,DC=bspWeb,DC=local"
```

**STEP 1.3: Display information containers organizational units. DSQUERY OU**

```
DSQUERY OU
C:\Windows\SYSVOL\sysvol\bspWeb.local>DSQUERY OU
"OU=Domain Controllers,DC=bspWeb,DC=local"
```

**STEP 1.4: Display server information domain. DSQUERY SERVER**

```
DSQUERY SERVER
C:\Windows\SYSVOL\sysvol\bspWeb.local>DSQUERY SERVER
"CN=SVRPRINC00,CN=Servers,CN=Default-First-Site-Name,CN=Sites,CN=Configuration,D
C=bspWeb,DC=local"
```

**STEP 1.5: Display information domain users. DSQUERY USER**

```
DSQUERY USER
C:\Windows\SYSVOL\sysvol\bspWeb.local>DSQUERY SERVER
"CN=SVRPRINC00,CN=Servers,CN=Default-First-Site-Name,CN=Sites,CN=Configuration,D
C=bspWeb,DC=local"

C:\Windows\SYSVOL\sysvol\bspWeb.local>DSQUERY  USER
"CN=Administrador,CN=Users,DC=bspWeb,DC=local"
"CN=Invitado,CN=Users,DC=bspWeb,DC=local"
"CN=krbtgt,CN=Users,DC=bspWeb,DC=local"
"CN=RUBEN,CN=Users,DC=bspWeb,DC=local"
"CN=ADRIAN,CN=Users,DC=bspWeb,DC=local"
"CN=JORGE,CN=Users,DC=bspWeb,DC=local"
"CN=ASER,CN=Users,DC=bspWeb,DC=local"
```

**STEP 1.6: Display information disk quotas and directories. DSQUERY QUOTA**

```
DSQUERY  QUOTA
DSQUERY  QUOTA  DOMAINROOT
```

**STEP 1.7: Display information by search sites in the Active Directory. DSQUERY SITE**

```
C:\Windows\system32> DSQUERY SITE
"CN=Default-First-Site-Name,CN=Sites,CN=Configuration,DC=dawprog0,DC=local"
```

**STEP 1.8: Find objects Active Directory partition. DSQUERY PARTITION**

```
C:\Windows\system32> DSQUERY PARTITION
"CN=Configuration,DC=dawprog0,DC=local"
"DC=dawprog0,DC=local"
"CN=Schema,CN=Configuration,DC=dawprog0,DC=local"
"DC=DomainDnsZones,DC=dawprog0,DC=local"
"DC=ForestDnsZones,DC=dawprog0,DC=local"
```

**STEP 1.10: Display by partition objects. DSQUERY PARTITION -PART DAW***

```
C:\Windows\system32>DSQUERY PARTITION -PART DAW*
"DC=dawprog0,DC=local"
```

**STEP 1.11: Display by domain. DSQUERY PARTITION -S DAWPROG0**

```
C:\Windows\system32>DSQUERY PARTITION -S DAWPROG0
"CN=Configuration,DC=dawprog0,DC=local"
"DC=dawprog0,DC=local"
"CN=Schema,CN=Configuration,DC=dawprog0,DC=local"
```

```
"DC=DomainDnsZones,DC=dawprog0,DC=local"
"DC=ForestDnsZones,DC=dawprog0,DC=local"
```

## STEP 1.12: Display by Server. DSQUERY PARTITION -S WIN-2OIUP9JNGLS

```
C:\Windows\system32>DSQUERY PARTITION -s WIN-2OIUP9JNGLS
"CN=Configuration,DC=dawprog0,DC=local"
"DC=dawprog0,DC=local"
"CN=Schema,CN=Configuration,DC=dawprog0,DC=local"
"DC=DomainDnsZones,DC=dawprog0,DC=local"
"DC=ForestDnsZones,DC=dawprog0,DC=local"
```

## STEP 2: SET environment variables

Environment variables can be reach:
- **Global:** The defined at the prompt with SET, or in any batch file.
- **Local:** are defined within a batch file between SETLOCAL and ENDLOCAL.

a) Ayuda variables de ambiente.
   SET /?
b) Display all variables.
   SET
c) Display all variables that begin with the letters indicated.
   SET A
   SET PA

```
C:\Windows\system32>SET A
ALLUSERSPROFILE=C:\ProgramData
APPDATA=C:\Users\Administrador\AppData\Roaming

C:\Windows\system32>SET PA
Path=C:\Windows\system32;C:\Windows;C:\Windows\System32\Wbem;C:\Windows\System32
\WindowsPowerShell\v1.0\
PATHEXT=.COM;.EXE;.BAT;.CMD;.VBS;.VBE;.JS;.JSE;.WSF;.WSH;.MSC
```

> **NOTA:** si se especifica las variables entre:
> SETLOCAL
> SET A=5
> ........
> ENDLOCAL
> son variables locales, sino son globales.

d) Use of variables in .bat files .CMD

   d.1) Local area. It is defined in the first lines of the file. From that moment all defined variables are local to the batch file it ends when it ends or destroyed the variable used. The file must end with ENDLOCAL.
   SETLOCAL

   d.2) The completion of the scope of local variables in a batch file. A variable is valid from SETLOCAL to ENDLOCAL.
   ENDLOCAL

   d.3) Create a variable or define a variable.
   SET nombre=valor

   d.4) Deleting a variable or destroy it.
   SET nombre=

   d.5) Read a variable from the keyboard. The value specified in equality is the string that appears in the question of application of the value. If no value is entered the variable does not exist.
   SET /P nombre=Cadena a visualizar o pregunta
   ```
   C:\Users\baldo>SET    /P    nombre=Cadena a visualizar o pregunta
   Cadena a visualizar o pregunta
   ```

   d.6) Operate with numeric variables:
   ```
   C:\Windows\system32>SET   a= 5

   C:\Windows\system32>SET   b=6

   C:\Windows\system32>SET    /A    resul=%a%  +   %b%
   11
   ```

   > If the assignment to a variable is correct a line is blank, then the PROMPT

   d.7) Read a system variable and replace its value.
   Acceder al directorio de trabajo del usuario.
   CD \%USERPROFILE%
   Visualizar el nombre del servidor.
   ECHO %LOGONSERVER%
   \\SVRPRINC00

   d.8) String variables using concatenation (SUM OF CHARACTERS).
   ```
   C:\Users\aprendiz>SET   DATA1=" HELLO "

   C:\Users\aprendiz>SET   DATA2="JUAN"

   C:\Users\aprendiz>SET   EXIST=%DATA1% %DATA2%

   C:\Users\aprendiz>ECHO      RESULT IS: %EXIST%
      RESULT IS: " HELLO " "JUAN"
   ```

# STEP 3: Environment variables stored in the registry or permanent. SETX

a) Variables stored in the registry or permanent.
   SETX

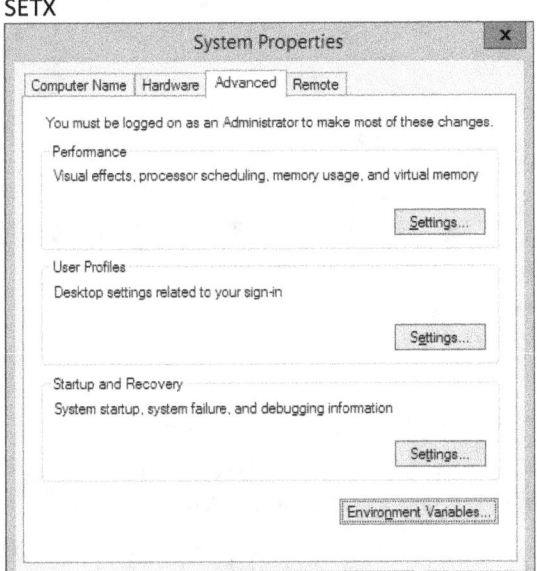

**Characteristics**
1) SETX writes the master environment variables in the registry.
2) In a local system, variables created or modified with this tool will be available in future command windows, but not in the current window CMD.exe commands.
3) In a remote system, variables created or modified with this tool will be available at the next logon.
4) The valid data types are registry key REG_DWORD, REG_EXPAND_SZ, REG_SZ, REG_MULTI_SZ.
5) Compatible Subtrees: HKEY_LOCAL_MACHINE (HKLM), HKEY_CURRENT_USER (HKCU).
6) The delimiters are case-sensitive.
7) REG_DWORD registry values are extracted in decimal format.

b) Environment Variables.

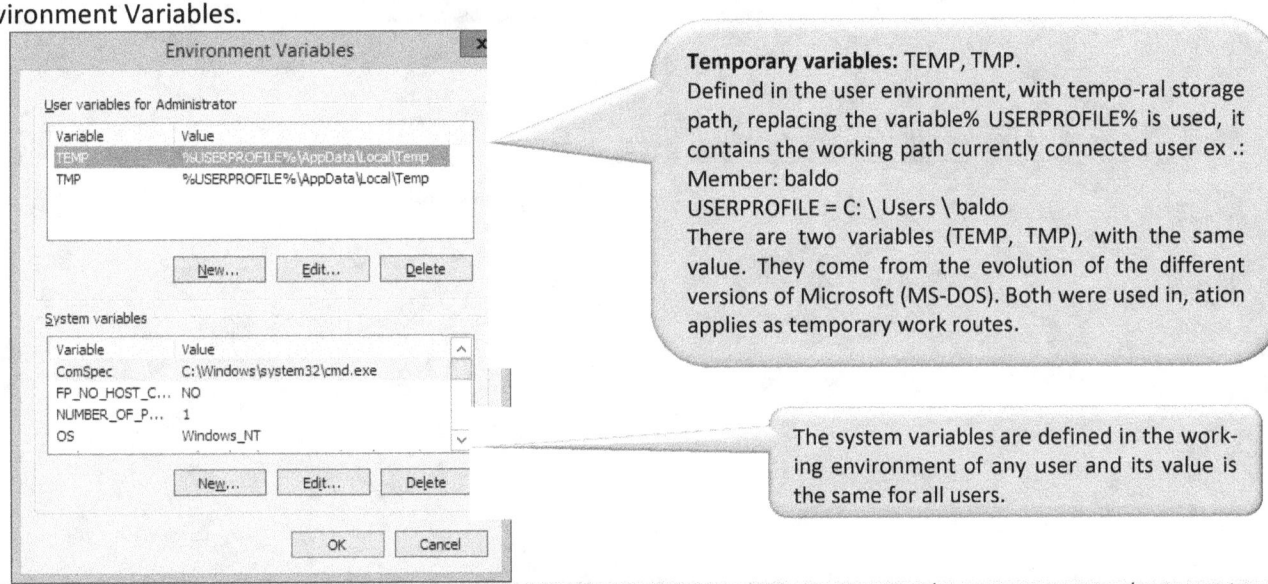

**Temporary variables:** TEMP, TMP.
Defined in the user environment, with tempo-ral storage path, replacing the variable% USERPROFILE% is used, it contains the working path currently connected user ex .:
Member: baldo
USERPROFILE = C: \ Users \ baldo
There are two variables (TEMP, TMP), with the same value. They come from the evolution of the different versions of Microsoft (MS-DOS). Both were used in, ation applies as temporary work routes.

The system variables are defined in the working environment of any user and its value is the same for all users.

```
C:\Windows\SYSVOL\sysvol\bspWeb.local>SETX /s SVRPRINC00 /U bspweb.local\Administrador /P Practi-
ca2017*  Variable-baldo "Ya estoy definiendo un valor en el servidor"  /M

SUCCESS: Specified value was saved..
```

Access the server.
    Start
        Run
            C:\> REGEDIT

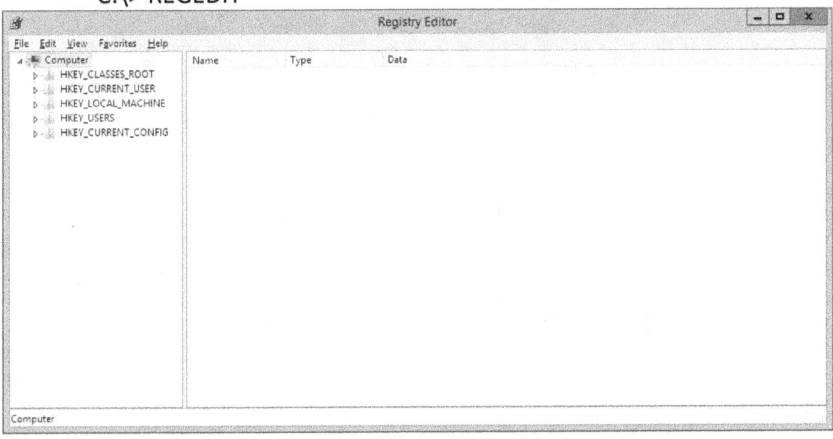

Edition

Buscar (ó F3)

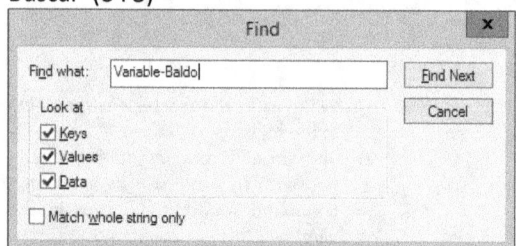

On the server we appear.

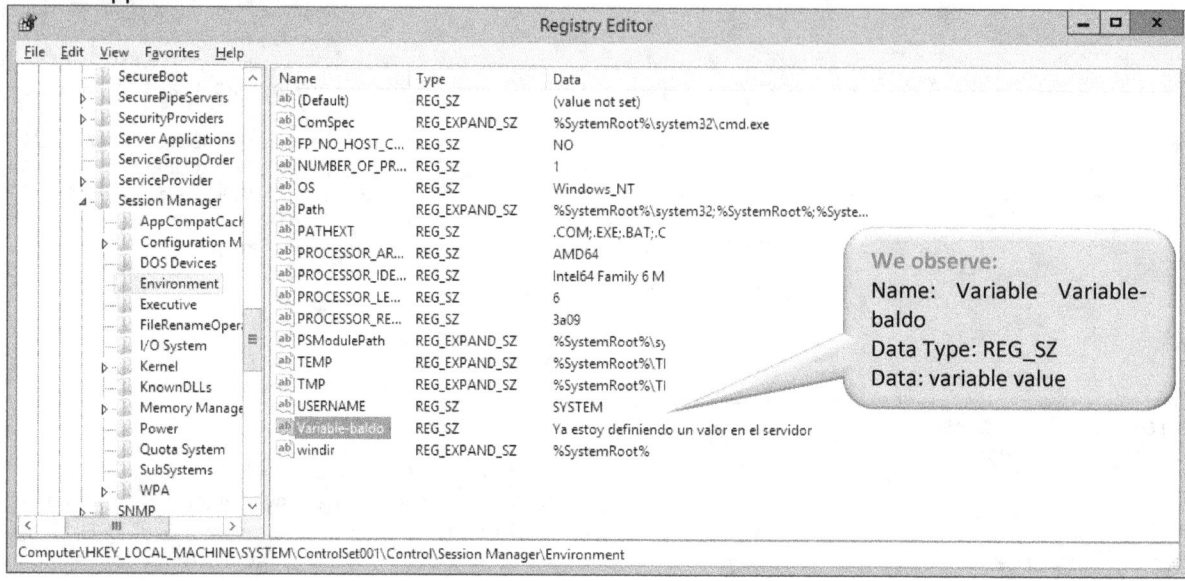

We observe:
Name: Variable Variable-baldo
Data Type: REG_SZ
Data: variable value

## STEP 1.9: Display by Output Format. DSQUERY PARTITION -O DN

```
C:\Windows\system32>DSQUERY PARTITION -O DN
"CN=Configuration,DC=dawprog0,DC=local"
"DC=dawprog0,DC=local"
"CN=Schema,CN=Configuration,DC=dawprog0,DC=local"
"DC=DomainDnsZones,DC=dawprog0,DC=local"
"DC=ForestDnsZones,DC=dawprog0,DC=local"

C:\Windows\system32>DSQUERY PARTITION -O RDN
"Enterprise Configuration"
"DAWPROG0"
"Enterprise Schema"
"ce9a0b60-1091-4c84-8c98-b21b3d4c15cc"
"dd995531-7dc2-4b21-a19d-a404e16bf405"
```

# PRACTICE 7: Display and set the partitions and quotas, in AD DS system.

**DESCRIPTION:**
There are two basic types of disk quotas.
- **Fee use or share blocks**, limits the amount of disk space that can be used.
- **Fee file or inode limits** the number of files and directories that can be created.

In addition, managers usually define a warning level, or soft quota, in which the user that they are approaching their limit Reportedly, which is less than the cash limit, or hard quota.

There may be a small grace interval, which allows users to temporarily violate their quotas by certain amounts, if necessary. When a soft quota is violated, the system normally sends the user (and sometimes the administrator also) some sort of message.

It is the system administrator who defines a use fee for a particular user or group.

## STEP 1: Display information disk quotas and directories. DSQUERY QUOTA
    DSQUERY QUOTA
    DSQUERY QUOTA DOMAINROOT

## STEP 2: System Information. FSUTIL
    FSUTIL

### STEP 2.1: Display volume information. FSUTIL FSINFO NTFSINFO

```
FSUTIL FSINFO NTFSINFO  C:
C:\Windows\SYSVOL\sysvol\bspWeb.local>FSUTIL     FSINFO NTFSINFO    C:
NTFS Volume Serial Number :       0x3a98c5d498c58eb7
NTFS Version     :                3.1
LFS Version      :                2.0
Number Sectors   :                0x0000000003f4ffff
Total Clusters   :                0x00000000007e9fff
Free Clusters    :                0x00000000005e9023
Total Reserved   :                0x0000000000020630
Bytes Per Sector       :          512
Bytes Per Physical Sector :       4096
Bytes Per Cluster      :          4096
Bytes Per FileRecord Segment    : 1024
Clusters Per FileRecord Segment : 0
Mft Valid Data Length  :          0x0000000005480000
Mft Start Lcn    :                0x00000000000c0000
Mft2 Start Lcn   :                0x0000000000000002
Mft Zone Start   :                0x00000000000c5480
Mft Zone End     :                0x00000000000cc820
Resource Manager Identifier :     EB425EDC-6687-11E6-8938-89BF2B7ECBDE
```

### STEP 2.2: Display volume information file system. FSUTIL FSINFO VOLUMEINFO

```
FSUTIL FSINFO VOLUMEINFO  C:
C:\Windows\SYSVOL\sysvol\bspWeb.local>FSTUTIL     FSINFO VOLUMEINFO    C:

Volume Name :
Volume Serial Number : 0x98c58eb7
Max Component Length : 255
File System Name : NTFS
Is ReadWrite
Supports Case-sensitive filenames
Preserves Case of filenames
Supports Unicode in filenames
Preserves & Enforces ACL's
Supports file-based Compression
Supports Disk Quotas
Supports Sparse files
Supports Reparse Points
Supports Object Identifiers
Supports Encrypted File System
Supports Named Streams
Supports Transactions
Supports Hard Links
Supports Extended Attributes
Supports Open By FileID
Supports USN Journal
```

### STEP 2.3: Establish a disk quota. FSUTIL QUOTA MODIFY
    FSUTIL QUOTA MODIFY warns  max-size   username
    FSUTIL QUOTA MODIFY C: 5242880 6291456 ASER

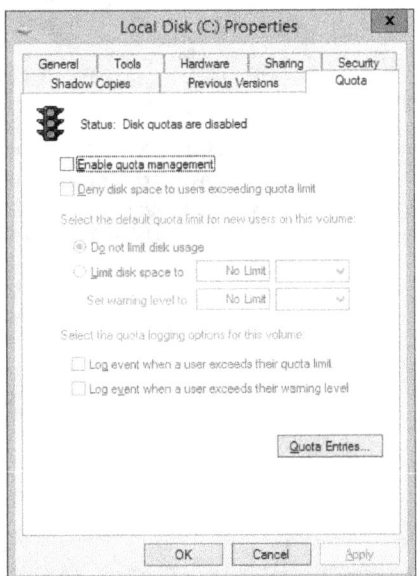
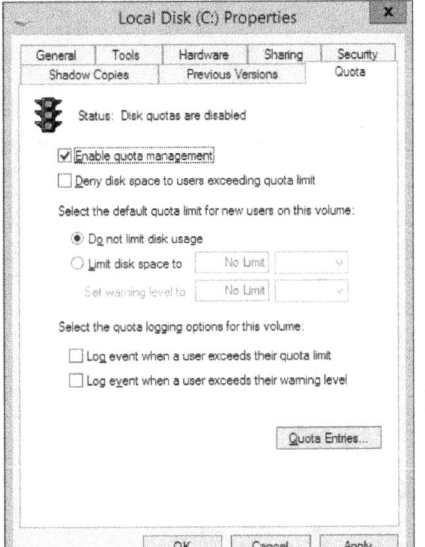

NOTE: Setting sharing quota for a particular user.

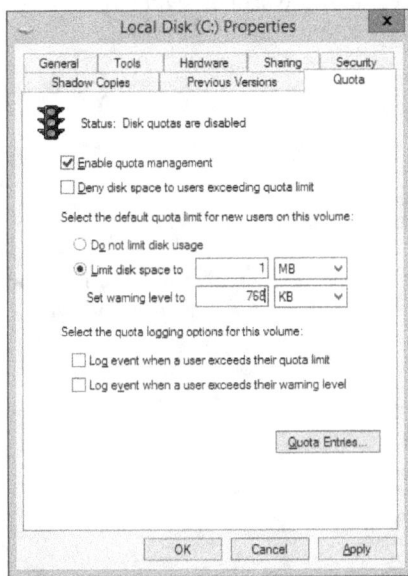

BUTTON quota values ...
　　The following window "Quota Entries for (C :)" appears

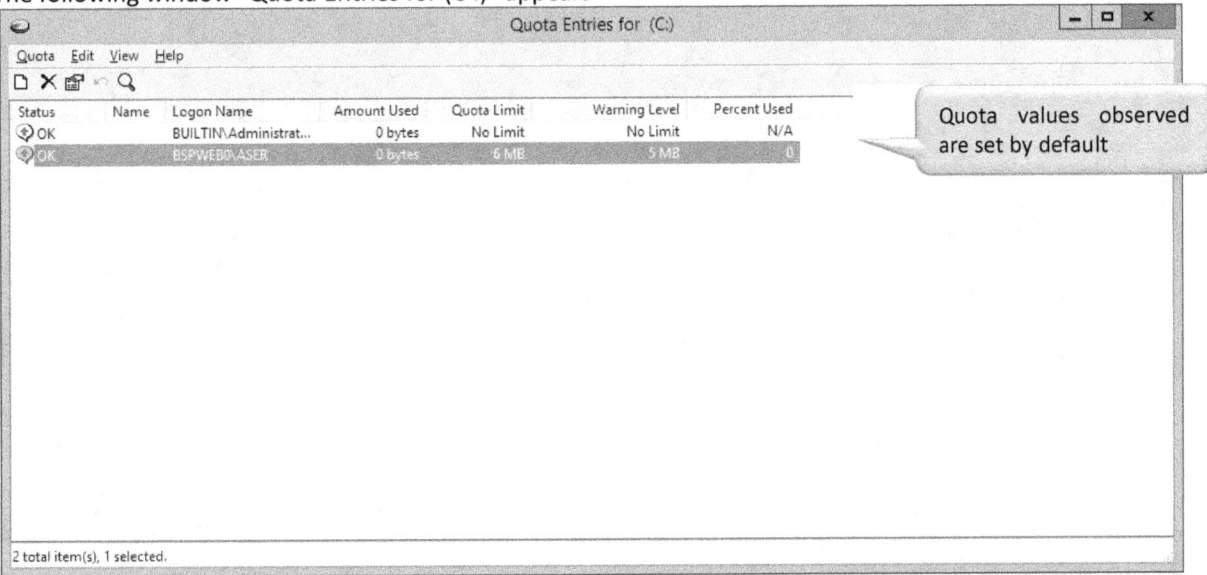

Quota values observed are set by default

　　To a particular user it is established ASER a maximum quota or quota limit to 6 Mbytes, and a warning level when it reaches 5 Mbytes
　　　　FSUTIL QUOTA MODIFY C: 5242880 6291456 ASER

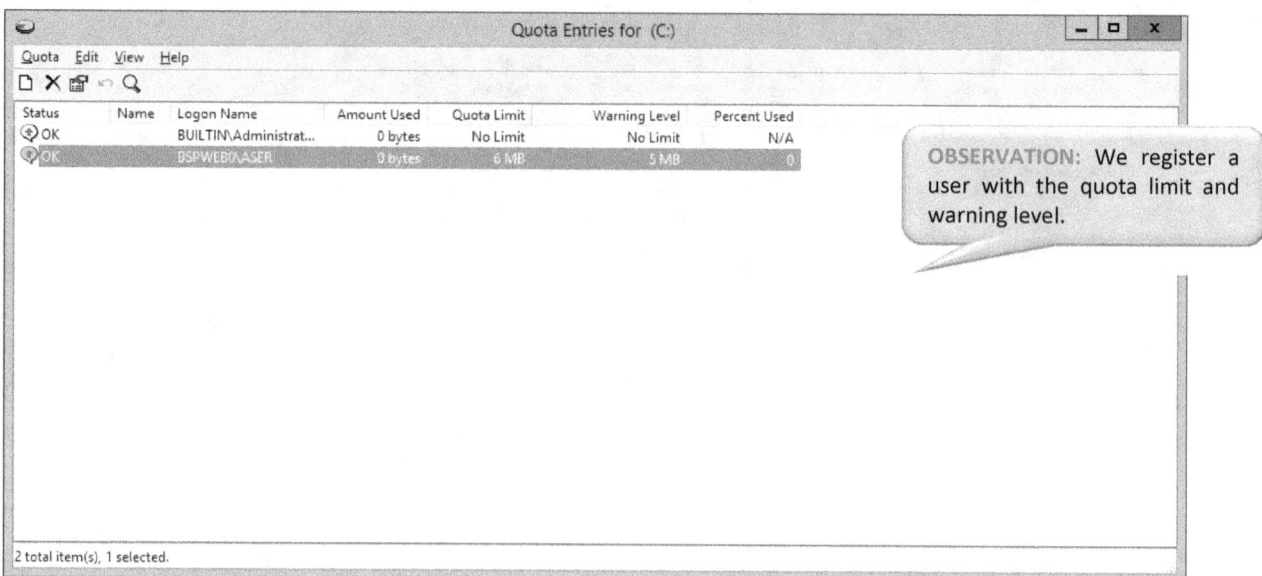

To a particular user ADRIAN set a maximum quota or quota limit to 3 Mbytes, and a warning level when you reach the 2.6 Mbytes.

FSUTIL QUOTA MODIFY C: 2726298 3145728 ADRIAN

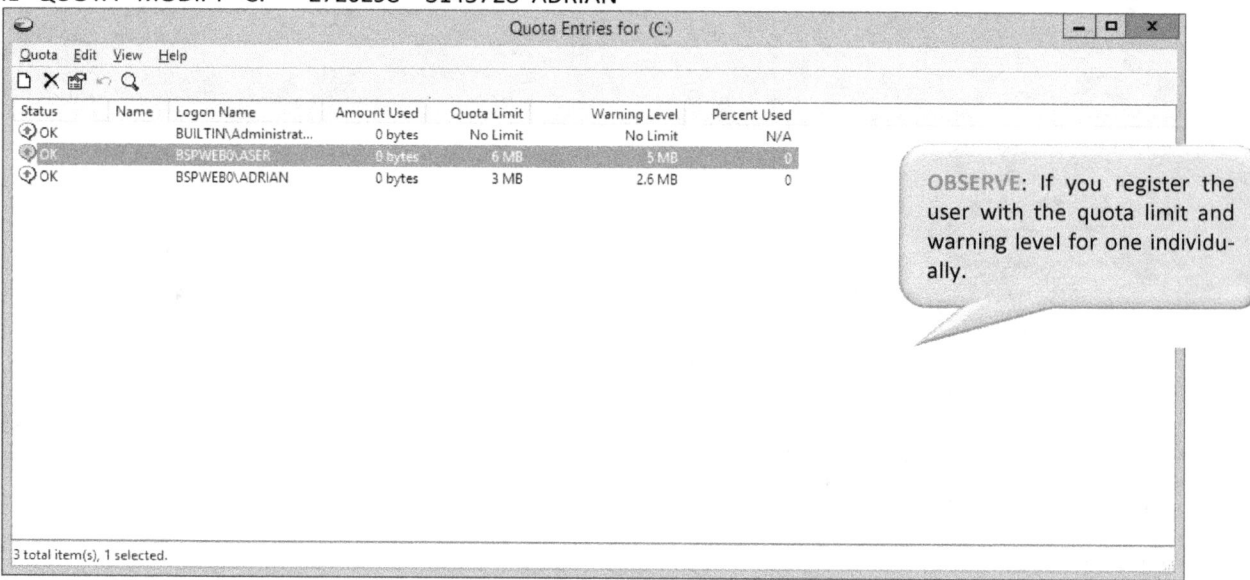

## PRACTICE 8: Restrictions Windows key level in AD DS.

**DESCRIPTION:**

NET ACCOUNTS updates the database and modifies user accounts password requirements and start-sion is for all accounts.

When used without options, NET ACCOUNTS displays the current settings for password, logon limits and in-training domain.

    NET ACCOUNTS

### STEP 1: Display lockout policies.

a) Display basic lockout policies.
```
C:\Windows\SYSVOL\sysvol\bspWeb.local>NET ACCOUNTS
Force user logoff how long after time expires?:    Never
Minimum password age (days):                       1
Maximum password age (days):                       42
Minimum password length:                           7
Length of password history maintained:             24
Lockout threshold:                                 Never
Lockout duration (minutes):                        30
Lockout observation window (minutes):              30
Computer role:                                     PRIMARY
The command completed successfully.
```

b) Set blocking force close to 15 min.
    NET ACCOUNTS /FORCELOGOFF:15 /DOMAIN
    NET ACCOUNTS

c) Stable time NEVER forced to close.
    NET ACCOUNTS /FORCELOGOFF:NO /DOMAIN
    NET ACCOUNTS

d) Set the minimum number of password characters will be 5.
    NET ACCOUNTS /MINPWLEN:5 /DOMAIN
    NET ACCOUNTS
    NET ACCOUNTS /MINPWLEN:NO /DOMAIN
    NET ACCOUNTS /HELP

> NET ACCOUNTS
> [/FORCELOGOFF:{minutos | NO}]
>     [/MINPWLEN:longitud]
>     [/MAXPWAGE:{días | UNLIMITED}]
>     [/MINPWAGE:días]
>     [/UNIQUEPW:número] [/DOMAIN]

You can get more help with command NET HELPMSG 3506.

e) Set the number of days that the key is valid. After the account expires.
    NET ACCOUNTS /MAXPWAGE: 60 /DOMAIN
    NET ACCOUNTS

f) The key never expires.
    NET ACCOUNTS /MAXPWAGE:UNLIMITED /DOMAIN
    NET ACCOUNTS

g) Number of days that lasts at least the key (default = 0).
    NET ACCOUNTS /MINPWAGE:40 /DOMAIN

h) Number of keys that are stored (24) does not allow repeated until the number of times are not changed.
    NET ACCOUNTS /UNIQUEPW:10 /DOMAIN

### STEP 2: Update policies. GPUPDATE

    GPUPDATE

a) Forcing upgrade, the directives.
    GPUPDATE /FORCE

> It is advisable to run once changed the values of the directives. Regardless of their change in command line or graphical environment.

### STEP 3: Display computer information directives. GPRESULT

    GPRESULT

a) Display normal summary.
    GPRESULT /R
```
C:\Windows\SYSVOL\sysvol\bspWeb.local>GPRESULT   /R

Microsoft (R) Windows (R) Operating System Group Policy Result tool v2.0
c 2013 Microsoft Corporation. All rights reserved.

Created on 9/2/2016 at 10:52:41 AM

RSOP data for BSPWEB0\Administrator on SVRPRINC00 : Logging Mode
-----------------------------------------------------------------

OS Configuration:            Primary Domain Controller
OS Version:                  6.3.9600
Site Name:                   Default-First-Site-Name
Roaming Profile:             N/A
Local Profile:               C:\Users\Administrator
Connected over a slow link?: No
```

```
COMPUTER SETTINGS
-----------------

    Last time Group Policy was applied: 9/2/2016 at 10:51:29 AM
    Group Policy was applied from:      SVRPRINC00.bspWeb.local
    Group Policy slow link threshold:   500 kbps
    Domain Name:                        BSPWEB0
    Domain Type:                        Windows 2008 or later

    Applied Group Policy Objects
    ----------------------------
        Default Domain Controllers Policy
        Default Domain Policy

    The following GPOs were not applied because they were filtered out
    ------------------------------------------------------------------
        Local Group Policy
            Filtering:  Not Applied (Empty)

    The computer is a part of the following security groups
    -------------------------------------------------------
        BUILTIN\Administrators
        Everyone
        BUILTIN\Users
        BUILTIN\Pre-Windows 2000 Compatible Access
        Windows Authorization Access Group
        NT AUTHORITY\NETWORK
        NT AUTHORITY\Authenticated Users
        This Organization
        SVR2017BSP$
        Domain Controllers
        NT AUTHORITY\ENTERPRISE DOMAIN CONTROLLERS
        Authentication authority asserted identity
        Denied RODC Password Replication Group
        System Mandatory Level


USER SETTINGS
-------------
    CN=Administrator,CN=Users,DC=bspWeb,DC=local
    Last time Group Policy was applied: 9/2/2016 at 8:25:38 AM
    Group Policy was applied from:      SVRPRINC00.bspWeb.local
    Group Policy slow link threshold:   500 kbps
    Domain Name:                        BSPWEB0
    Domain Type:                        Windows 2008 or later

    Applied Group Policy Objects
    ----------------------------
        N/A

    The following GPOs were not applied because they were filtered out
    ------------------------------------------------------------------
        Local Group Policy
            Filtering:  Not Applied (Empty)

    The user is a part of the following security groups
    ---------------------------------------------------
        Domain Users
        Everyone
        BUILTIN\Administrators
        Remote Desktop Users
        BUILTIN\Users
        BUILTIN\Pre-Windows 2000 Compatible Access
        REMOTE INTERACTIVE LOGON
        NT AUTHORITY\INTERACTIVE
        NT AUTHORITY\Authenticated Users
        This Organization
        LOCAL
        Domain Admins
        Group Policy Creator Owners
        Enterprise Admins
        Schema Admins
        Authentication authority asserted identity
        Denied RODC Password Replication Group
        High Mandatory Level
```

**b) Display an extended summary. GPRESULT /Z**

```
C:\Windows\SYSVOL\sysvol\bspWeb.local>GPRESULT    /Z
Microsoft (R) Windows (R) Operating System Group Policy Result tool
v2.0
c 2013 Microsoft Corporation. All rights reserved.

Created on 9/2/2016 at 10:54:58 AM

RSOP data for BSPWEB0\Administrator on SVRPRINC00 : Logging Mode
----------------------------------------------------------------

OS Configuration:            Primary Domain Controller
OS Version:                  6.3.9600
Site Name:                   Default-First-Site-Name
Roaming Profile:             N/A
Local Profile:               C:\Users\Administrator
Connected over a slow link?: No

COMPUTER SETTINGS
-----------------

    Last time Group Policy was applied: 9/2/2016 at 10:51:29 AM
    Group Policy was applied from:      SVRPRINC00.bspWeb.local
    Group Policy slow link threshold:   500 kbps
    Domain Name:                        BSPWEB0
    Domain Type:                        Windows 2008 or later

    Applied Group Policy Objects
    ----------------------------
        Default Domain Controllers Policy
        Default Domain Policy

    The following GPOs were not applied because they were filtered out
    ------------------------------------------------------------------
        Local Group Policy
            Filtering:  Not Applied (Empty)

    The computer is a part of the following security groups
    -------------------------------------------------------
        BUILTIN\Administrators
        Everyone
        BUILTIN\Users
        BUILTIN\Pre-Windows 2000 Compatible Access
        Windows Authorization Access Group
        NT AUTHORITY\NETWORK
        NT AUTHORITY\Authenticated Users
        This Organization
        SVR2017BSP$
        Domain Controllers
        NT AUTHORITY\ENTERPRISE DOMAIN CONTROLLERS
        Authentication authority asserted identity
        Denied RODC Password Replication Group
        System Mandatory Level

    Resultant Set Of Policies for Computer
    --------------------------------------

        Software Installations
        ----------------------
            N/A

        Startup Scripts
        ---------------
            N/A
```

# PRACTICAL BOOKLET 1: Commands Windows network and AD DS

```
Shutdown Scripts
----------------
    N/A

Account Policies
----------------
    GPO: Default Domain Policy
        Policy:            MaxRenewAge
        Computer Setting:  7

    GPO: Default Domain Policy
        Policy:            MaximumPasswordAge
        Computer Setting:  42

    GPO: Default Domain Policy
        Policy:            MinimumPasswordAge
        Computer Setting:  1

    GPO: Default Domain Policy
        Policy:            MaxServiceAge
        Computer Setting:  600

    GPO: Default Domain Policy
        Policy:            LockoutBadCount
        Computer Setting:  N/A

    GPO: Default Domain Policy
        Policy:            MaxClockSkew
        Computer Setting:  5

    GPO: Default Domain Policy
        Policy:            MaxTicketAge
        Computer Setting:  10

    GPO: Default Domain Policy
        Policy:            PasswordHistorySize
        Computer Setting:  24

    GPO: Default Domain Policy
        Policy:            MinimumPasswordLength
        Computer Setting:  7

Audit Policy
------------
    N/A

User Rights
-----------
    GPO: Default Domain Controllers Policy
        Policy:            MachineAccountPrivilege
        Computer Setting:  Authenticated Users

    GPO: Default Domain Controllers Policy
        Policy:            ChangeNotifyPrivilege
        Computer Setting:  Everyone
                           LOCAL SERVICE
                           NETWORK SERVICE
                           Administrators
                           Window Manager\Window Manager Group
                           Authenticated Users
                           Pre-Windows 2000 Compatible Access

    GPO: Default Domain Controllers Policy
        Policy:            IncreaseBasePriorityPrivilege
        Computer Setting:  Administrators

    GPO: Default Domain Controllers Policy
        Policy:            TakeOwnershipPrivilege
        Computer Setting:  Administrators

    GPO: Default Domain Controllers Policy
        Policy:            RestorePrivilege
        Computer Setting:  Administrators
                           Backup Operators
                           Server Operators

    GPO: Default Domain Controllers Policy
        Policy:            DebugPrivilege
        Computer Setting:  Administrators

    GPO: Default Domain Controllers Policy
        Policy:            SystemTimePrivilege
        Computer Setting:  LOCAL SERVICE
                           Administrators
                           Server Operators

    GPO: Default Domain Controllers Policy
        Policy:            SecurityPrivilege
        Computer Setting:  Administrators

    GPO: Default Domain Controllers Policy
        Policy:            ShutdownPrivilege
        Computer Setting:  Administrators
                           Backup Operators
                           Server Operators
                           Print Operators

    GPO: Default Domain Controllers Policy
        Policy:            AuditPrivilege
        Computer Setting:  LOCAL SERVICE
                           NETWORK SERVICE

    GPO: Default Domain Controllers Policy
        Policy:            InteractiveLogonRight
        Computer Setting:  Administrators
                           Backup Operators
                           Account Operators
                           Server Operators
                           Print Operators
                           ENTERPRISE DOMAIN CONTROLLERS

    GPO: Default Domain Controllers Policy
        Policy:            CreatePagefilePrivilege
        Computer Setting:  Administrators

    GPO: Default Domain Controllers Policy
        Policy:            BatchLogonRight
        Computer Setting:  Administrators
                           Backup Operators
                           Performance Log Users

    GPO: Default Domain Controllers Policy
        Policy:            NetworkLogonRight
        Computer Setting:  Everyone
                           Administrators
                           Authenticated Users
                           ENTERPRISE DOMAIN CONTROLLERS
                           Pre-Windows 2000 Compatible Access

    GPO: Default Domain Controllers Policy
        Policy:            SystemProfilePrivilege
        Computer Setting:  Administrators
                           NT SERVICE\WdiServiceHost

    GPO: Default Domain Controllers Policy
        Policy:            RemoteShutdownPrivilege
        Computer Setting:  Administrators
                           Server Operators

    GPO: Default Domain Controllers Policy
        Policy:            BackupPrivilege
        Computer Setting:  Administrators
                           Backup Operators
                           Server Operators

    GPO: Default Domain Controllers Policy
        Policy:            EnableDelegationPrivilege
        Computer Setting:  Administrators

    GPO: Default Domain Controllers Policy
        Policy:            UndockPrivilege
        Computer Setting:  Administrators

    GPO: Default Domain Controllers Policy
        Policy:            SystemEnvironmentPrivilege
        Computer Setting:  Administrators

    GPO: Default Domain Controllers Policy
        Policy:            LoadDriverPrivilege
        Computer Setting:  Administrators
                           Print Operators

    GPO: Default Domain Controllers Policy
        Policy:            IncreaseQuotaPrivilege
        Computer Setting:  LOCAL SERVICE
                           NETWORK SERVICE
                           Administrators

    GPO: Default Domain Controllers Policy
        Policy:            ProfileSingleProcessPrivilege
        Computer Setting:  Administrators

    GPO: Default Domain Controllers Policy
        Policy:            AssignPrimaryTokenPrivilege
        Computer Setting:  LOCAL SERVICE
                           NETWORK SERVICE

Security Options
----------------
    GPO: Default Domain Policy
        Policy:            PasswordComplexity
        Computer Setting:  Enabled

    GPO: Default Domain Policy
        Policy:            ClearTextPassword
        Computer Setting:  Not Enabled

    GPO: Default Domain Policy
        Policy:            ForceLogoffWhenHourExpire
        Computer Setting:  Not Enabled

    GPO: Default Domain Policy
        Policy:            RequireLogonToChangePassword
        Computer Setting:  Not Enabled

    GPO: Default Domain Policy
        Policy:            LSAAnonymousNameLookup
        Computer Setting:  Not Enabled

    GPO: Default Domain Policy
        Policy:            TicketValidateClient
        Computer Setting:  Enabled

    GPO: Default Domain Controllers Policy
        Policy:            @wsecedit.dll,-59013
        ValueName:         MA-
CHINE\System\CurrentControlSet\Services\NTDS\Parameters\LDAPServerInte
grity
        Computer Setting:  1

    GPO: Default Domain Controllers Policy
        Policy:            @wsecedit.dll,-59043
        ValueName:         MA-
CHINE\System\CurrentControlSet\Services\LanManServer\Parameters\Requir
eSecuritySignature
        Computer Setting:  1

    GPO: Default Domain Controllers Policy
        Policy:            @wsecedit.dll,-59044
```

```
              ValueName:         MA-
CHINE\System\CurrentControlSet\Services\LanManServer\Parameters\Enable
SecuritySignature
              Computer Setting:  1

        GPO: Default Domain Policy
              Policy:            @wsecedit.dll,-59058
              ValueName:         MA-
CHINE\System\CurrentControlSet\Control\Lsa\NoLMHash
              Computer Setting:  1

        GPO: Default Domain Controllers Policy
              Policy:            @wsecedit.dll,-59018
              ValueName:         MA-
CHINE\System\CurrentControlSet\Services\Netlogon\Parameters\RequireSig
nOrSeal
              Computer Setting:  1

        N/A

    Event Log Settings
    ------------------
        N/A

    Restricted Groups
    -----------------
        N/A

    System Services
    ---------------
        N/A

    Registry Settings
    -----------------
        N/A

    File System Settings
    --------------------
        N/A

    Public Key Policies
    -------------------
        N/A

    Administrative Templates
    ------------------------
        N/A

USER SETTINGS
-------------
    CN=Administrator,CN=Users,DC=bspWeb,DC=local
    Last time Group Policy was applied: 9/2/2016 at 8:25:38 AM
    Group Policy was applied from:      SVRPRINC00.bspWeb.local
    Group Policy slow link threshold:   500 kbps
    Domain Name:                        BSPWEB0
    Domain Type:                        Windows 2008 or later

    Applied Group Policy Objects
    ----------------------------
        N/A

    The following GPOs were not applied because they were filtered out
    ------------------------------------------------------------------

        Local Group Policy
            Filtering:  Not Applied (Empty)

    The user is a part of the following security groups
    ---------------------------------------------------
        Domain Users
        Everyone
        BUILTIN\Administrators
        Remote Desktop Users
        BUILTIN\Users
        BUILTIN\Pre-Windows 2000 Compatible Access
        REMOTE INTERACTIVE LOGON
        NT AUTHORITY\INTERACTIVE
        NT AUTHORITY\Authenticated Users
        This Organization
        LOCAL
        Domain Admins
        Group Policy Creator Owners
        Enterprise Admins
        Schema Admins
        Authentication authority asserted identity
        Denied RODC Password Replication Group
        High Mandatory Level

    The user has the following security privileges
    ----------------------------------------------
        Bypass traverse checking
        Manage auditing and security log
        Back up files and directories
        Restore files and directories
        Change the system time
        Shut down the system
        Force shutdown from a remote system
        Take ownership of files or other objects
        Debug programs
        Modify firmware environment values
        Profile system performance
        Profile single process
        Increase scheduling priority
        Load and unload device drivers
        Create a pagefile
        Adjust memory quotas for a process
        Remove computer from docking station
        Perform volume maintenance tasks
        Impersonate a client after authentication
        Create global objects
        Change the time zone
        Create symbolic links
        Enable computer and user accounts to be trusted for delegation
        Increase a process working set
        Add workstations to domain

    Resultant Set Of Policies for User
    ----------------------------------

        Software Installations
        ----------------------
            N/A

        Logon Scripts
        -------------
            N/A

        Logoff Scripts
        --------------
            N/A

        Public Key Policies
        -------------------
            N/A

        Administrative Templates
        ------------------------
            N/A

        Folder Redirection
        ------------------
            N/A

        Internet Explorer Browser User Interface
        ----------------------------------------
            N/A

        Internet Explorer Connection
        ----------------------------
            N/A

        Internet Explorer URLs
        ----------------------
            N/A

        Internet Explorer Security
        --------------------------
            N/A

        Internet Explorer Programs
        --------------------------
            N/A
```

c) Display level information of a user field.
    GPRESULT /SCOPE /USER --> nota ?
d) Display system information in a domain and a user, display detail.
    d.1) From a Windows 10 Professional team.
    d.2) From a domain server.

**C:\Windows\system32>gpresult /S SVR-BSP-00 /u bspWeb.local\Administrador /P Practica2017* /V**

WARNING: Ignoring the user credentials for the local system.

```
C:\Windows\SYSVOL\sysvol\bspWeb.local>gpresult /S SVRPRINC00 /u bspWeb.local\Administrador /P
Practica2017* /V

Microsoft (R) Windows (R) Operating System Group Policy Result tool
v2.0
c 2013 Microsoft Corporation. All rights reserved.

Created on 9/2/2016 at 10:59:40 AM

RSOP data for BSPWEB0\Administrator on SVRPRINC00 : Logging Mode
-----------------------------------------------------------------

OS Configuration:           Primary Domain Controller
OS Version:                 6.3.9600
Site Name:                  Default-First-Site-Name
Roaming Profile:            N/A
Local Profile:              C:\Users\Administrator
Connected over a slow link?: No

COMPUTER SETTINGS
-----------------
    Last time Group Policy was applied: 9/2/2016 at 10:56:29 AM
    Group Policy was applied from:      SVRPRINC00.bspWeb.local
    Group Policy slow link threshold:   500 kbps
    Domain Name:                        BSPWEB0
    Domain Type:                        Windows 2008 or later
```

# PRACTICAL BOOKLET 1: Commands Windows network and AD DS

```
Applied Group Policy Objects
-----------------------------
    Default Domain Controllers Policy
    Default Domain Policy

The following GPOs were not applied because they were filtered out
------------------------------------------------------------------
    Local Group Policy
        Filtering:  Not Applied (Empty)

The computer is a part of the following security groups
-------------------------------------------------------
    BUILTIN\Administrators
    Everyone
    BUILTIN\Users
    BUILTIN\Pre-Windows 2000 Compatible Access
    Windows Authorization Access Group
    NT AUTHORITY\NETWORK
    NT AUTHORITY\Authenticated Users
    This Organization
    SVR2017BSP$
    Domain Controllers
    NT AUTHORITY\ENTERPRISE DOMAIN CONTROLLERS
    Authentication authority asserted identity
    Denied RODC Password Replication Group
    System Mandatory Level

Resultant Set Of Policies for Computer
--------------------------------------
    Software Installations
    ----------------------
        N/A

    Startup Scripts
    ---------------
        N/A

    Shutdown Scripts
    ----------------
        N/A

    Account Policies
    ----------------
        GPO: Default Domain Policy
            Policy:            MaxRenewAge
            Computer Setting:  7

        GPO: Default Domain Policy
            Policy:            MaximumPasswordAge
            Computer Setting:  42

        GPO: Default Domain Policy
            Policy:            MinimumPasswordAge
            Computer Setting:  1

        GPO: Default Domain Policy
            Policy:            MaxServiceAge
            Computer Setting:  600

        GPO: Default Domain Policy
            Policy:            LockoutBadCount
            Computer Setting:  N/A

        GPO: Default Domain Policy
            Policy:            MaxClockSkew
            Computer Setting:  5

        GPO: Default Domain Policy
            Policy:            MaxTicketAge
            Computer Setting:  10

        GPO: Default Domain Policy
            Policy:            PasswordHistorySize
            Computer Setting:  24

        GPO: Default Domain Policy
            Policy:            MinimumPasswordLength
            Computer Setting:  7

    Audit Policy
    ------------
        N/A

    User Rights
    -----------
        GPO: Default Domain Controllers Policy
            Policy:            MachineAccountPrivilege
            Computer Setting:  Authenticated Users

        GPO: Default Domain Controllers Policy
            Policy:            ChangeNotifyPrivilege
            Computer Setting:  Everyone
                               LOCAL SERVICE
                               NETWORK SERVICE
                               Administrators
                               Window Manager\Window Manager Group
                               Authenticated Users
                               Pre-Windows 2000 Compatible Access

        GPO: Default Domain Controllers Policy
            Policy:            IncreaseBasePriorityPrivilege
            Computer Setting:  Administrators

        GPO: Default Domain Controllers Policy
            Policy:            TakeOwnershipPrivilege
            Computer Setting:  Administrators

        GPO: Default Domain Controllers Policy
            Policy:            RestorePrivilege
                               Computer Setting: Administrators
                               Backup Operators
                               Server Operators

        GPO: Default Domain Controllers Policy
            Policy:            DebugPrivilege
            Computer Setting:  Administrators

        GPO: Default Domain Controllers Policy
            Policy:            SystemTimePrivilege
            Computer Setting:  LOCAL SERVICE
                               Administrators
                               Server Operators

        GPO: Default Domain Controllers Policy
            Policy:            SecurityPrivilege
            Computer Setting:  Administrators

        GPO: Default Domain Controllers Policy
            Policy:            ShutdownPrivilege
            Computer Setting:  Administrators
                               Backup Operators
                               Server Operators
                               Print Operators

        GPO: Default Domain Controllers Policy
            Policy:            AuditPrivilege
            Computer Setting:  LOCAL SERVICE
                               NETWORK SERVICE

        GPO: Default Domain Controllers Policy
            Policy:            InteractiveLogonRight
            Computer Setting:  Administrators
                               Backup Operators
                               Account Operators
                               Server Operators
                               Print Operators
                               ENTERPRISE DOMAIN CONTROLLERS

        GPO: Default Domain Controllers Policy
            Policy:            CreatePagefilePrivilege
            Computer Setting:  Administrators

        GPO: Default Domain Controllers Policy
            Policy:            BatchLogonRight
            Computer Setting:  Administrators
                               Backup Operators
                               Performance Log Users

        GPO: Default Domain Controllers Policy
            Policy:            NetworkLogonRight
            Computer Setting:  Everyone
                               Administrators
                               Authenticated Users
                               ENTERPRISE DOMAIN CONTROLLERS
                               Pre-Windows 2000 Compatible Access

        GPO: Default Domain Controllers Policy
            Policy:            SystemProfilePrivilege
            Computer Setting:  Administrators
                               NT SERVICE\WdiServiceHost

        GPO: Default Domain Controllers Policy
            Policy:            RemoteShutdownPrivilege
            Computer Setting:  Administrators
                               Server Operators

        GPO: Default Domain Controllers Policy
            Policy:            BackupPrivilege
            Computer Setting:  Administrators
                               Backup Operators
                               Server Operators

        GPO: Default Domain Controllers Policy
            Policy:            EnableDelegationPrivilege
            Computer Setting:  Administrators

        GPO: Default Domain Controllers Policy
            Policy:            UndockPrivilege
            Computer Setting:  Administrators

        GPO: Default Domain Controllers Policy
            Policy:            SystemEnvironmentPrivilege
            Computer Setting:  Administrators

        GPO: Default Domain Controllers Policy
            Policy:            LoadDriverPrivilege
            Computer Setting:  Administrators
                               Print Operators

        GPO: Default Domain Controllers Policy
            Policy:            IncreaseQuotaPrivilege
            Computer Setting:  LOCAL SERVICE
                               NETWORK SERVICE
                               Administrators

        GPO: Default Domain Controllers Policy
            Policy:            ProfileSingleProcessPrivilege
            Computer Setting:  Administrators

        GPO: Default Domain Controllers Policy
            Policy:            AssignPrimaryTokenPrivilege
            Computer Setting:  LOCAL SERVICE
                               NETWORK SERVICE

    Security Options
    ----------------
        GPO: Default Domain Policy
            Policy:            PasswordComplexity
            Computer Setting:  Enabled
```

```
        GPO: Default Domain Policy
            Policy:            ClearTextPassword
            Computer Setting:  Not Enabled

        GPO: Default Domain Policy
            Policy:            ForceLogoffWhenHourExpire
            Computer Setting:  Not Enabled

        GPO: Default Domain Policy
            Policy:            RequireLogonToChangePassword
            Computer Setting:  Not Enabled

        GPO: Default Domain Policy
            Policy:            LSAAnonymousNameLookup
            Computer Setting:  Not Enabled

        GPO: Default Domain Policy
            Policy:            TicketValidateClient
            Computer Setting:  Enabled

        GPO: Default Domain Controllers Policy
            Policy:            @wsecedit.dll,-59013
            ValueName:         MA-
CHINE\System\CurrentControlSet\Services\NTDS\Parameters\LDAPServerInte
grity
            Computer Setting:  1

        GPO: Default Domain Controllers Policy
            Policy:            @wsecedit.dll,-59043
            ValueName:         MA-
CHINE\System\CurrentControlSet\Services\LanManServer\Parameters\Requir
eSecuritySignature
            Computer Setting:  1

        GPO: Default Domain Controllers Policy
            Policy:            @wsecedit.dll,-59044
            ValueName:         MA-
CHINE\System\CurrentControlSet\Services\LanManServer\Parameters\Enable
SecuritySignature
            Computer Setting:  1

        GPO: Default Domain Policy
            Policy:            @wsecedit.dll,-59058
            ValueName:         MA-
CHINE\System\CurrentControlSet\Control\Lsa\NoLMHash
            Computer Setting:  1

        GPO: Default Domain Controllers Policy
            Policy:            @wsecedit.dll,-59018
            ValueName:         MA-
CHINE\System\CurrentControlSet\Services\Netlogon\Parameters\RequireSig
nOrSeal
            Computer Setting:  1

        N/A

    Event Log Settings
    ------------------
        N/A

    Restricted Groups
    -----------------
        N/A

    System Services
    ---------------
        N/A

    Registry Settings
    -----------------
        N/A

    File System Settings
    --------------------
        N/A

    Public Key Policies
    -------------------
        N/A

    Administrative Templates
    ------------------------
        N/A

USER SETTINGS
-------------
    CN=Administrator,CN=Users,DC=bspWeb,DC=local
    Last time Group Policy was applied: 9/2/2016 at 8:25:38 AM
    Group Policy was applied from:      SVRPRINC00.bspWeb.local
    Group Policy slow link threshold:   500 kbps
    Domain Name:                        BSPWEB0
    Domain Type:                        Windows 2008 or later

    Applied Group Policy Objects
    ----------------------------
        N/A

    The following GPOs were not applied because they were filtered out
    ------------------------------------------------------------------
        Local Group Policy
            Filtering:  Not Applied (Empty)

    The user is a part of the following security groups
    ---------------------------------------------------
        Domain Users
        Everyone
        BUILTIN\Administrators
        Remote Desktop Users
        BUILTIN\Users
        BUILTIN\Pre-Windows 2000 Compatible Access
        REMOTE INTERACTIVE LOGON
        NT AUTHORITY\INTERACTIVE
        NT AUTHORITY\Authenticated Users
        This Organization
        LOCAL
        Domain Admins
        Group Policy Creator Owners
        Enterprise Admins
        Schema Admins
        Authentication authority asserted identity
        Denied RODC Password Replication Group
        High Mandatory Level

    The user has the following security privileges
    ----------------------------------------------
        Bypass traverse checking
        Manage auditing and security log
        Back up files and directories
        Restore files and directories
        Change the system time
        Shut down the system
        Force shutdown from a remote system
        Take ownership of files or other objects
        Debug programs
        Modify firmware environment values
        Profile system performance
        Profile single process
        Increase scheduling priority
        Load and unload device drivers
        Create a pagefile
        Adjust memory quotas for a process
        Remove computer from docking station
        Perform volume maintenance tasks
        Impersonate a client after authentication
        Create global objects
        Change the time zone
        Create symbolic links
        Enable computer and user accounts to be trusted for delegation
        Increase a process working set
        Add workstations to domain

    Resultant Set Of Policies for User
    ----------------------------------
        Software Installations
        ----------------------
            N/A

        Logon Scripts
        -------------
            N/A

        Logoff Scripts
        --------------
            N/A

        Public Key Policies
        -------------------
            N/A

        Administrative Templates
        ------------------------
            N/A

        Folder Redirection
        ------------------
            N/A

        Internet Explorer Browser User Interface
        ----------------------------------------
            N/A

        Internet Explorer Connection
        ----------------------------
            N/A

        Internet Explorer URLs
        ----------------------
            N/A

        Internet Explorer Security
        --------------------------
            N/A

        Internet Explorer Programs
        --------------------------
            N/A
```

# PRACTICE 9: Analyze and see the protocols and processes running sessions.

**DESCRIPTION:**

The session layer or session layer is the fifth level of the OSI model, which provides the mechanisms to control the dialogue between end systems applications may be dispensable although some applications use is mandatory.

The session layer provides the following services:
- **Control Dialogue:** (full-duplex, half-duplex).
- **Grouping:** The data stream can be marked to define data sets.
- **Recovery:** The session layer can provide a method of checkpoints, so that if some kind of failure between checkpoints occurs, the session entity can retransmit all data from the last checkpoint and not from the beginning.

## Layer protocols session

- **Protocol RCP (Remote Procedure Call)** is a protocol that allows a computer program to run code on a remote machine without having to worry about communications between them. The protocol is a major advancement over sockets used so far. RPCs are widely used within the client-server paradigm. It is the customer who initiates the process by asking the server running certain procedure or function and sending it back the result of the operation to the customer. Today is using XML as the IDL language to define and HTTP as the network protocol,
- **SCP (simple communication protocol):** The SCP protocol is basically identical to the PCR protocol unlike this, the data are encrypted during transfer, to avoid potential sniffers extract useful information packet of data packets. However, the protocol itself does not provide authentication and security; but it expects the underlying protocol, SSH, to secure it.
- **ASP (Session protocol APPLE TALK):** It was developed by Apple Computers, provides session establishment, maintenance and dismantling, as well as the sequence request. ASP is an intermediate protocol that relies on top of AppleTalk Transaction Protocol (ATP), which is the reliable original session level protocol Apple-Talk.

Ilustración 3. http://www.ipref.info/2009_08_01_archive.html, IP Reference Themes CCNA, CCNP and CCIE Routing & Switching, in August 2009. Complete Graph on session protocols TCP / IP 31.

## STEP 1: Protocol session. NETSTAT

We reported all active incoming and outgoing connections to our equipment, including data such as network protocols used, IP addresses and ports, connection status among others.

a) Displays all connections and ports listened.

```
C:\Windows\system32>netstat -a
Active Connections

  Proto  Local Address          Foreign Address        State
  TCP    0.0.0.0:42             SVRPRINC00:0           LISTENING
  TCP    0.0.0.0:80             SVRPRINC00:0           LISTENING
  . . . . . . . . . . . . . .
  TCP    0.0.0.0:49207          SVRPRINC00:0           LISTENING
  TCP    0.0.0.0:49222          SVRPRINC00:0           LISTENING
  TCP    0.0.0.0:63163          SVRPRINC00:0           LISTENING
  TCP    127.0.0.1:53           SVRPRINC00:0           LISTENING
  TCP    127.0.0.1:389          SVRPRINC00:49163       ESTABLISHED
  TCP    127.0.0.1:389          SVRPRINC00:49165       ESTABLISHED
  TCP    127.0.0.1:389          SVRPRINC00:62961       ESTABLISHED
  TCP    127.0.0.1:49163        SVRPRINC00:ldap        ESTABLISHED
  TCP    127.0.0.1:49165        SVRPRINC00:ldap        ESTABLISHED
  TCP    127.0.0.1:62961        SVRPRINC00:ldap        ESTABLISHED
```

```
TCP    192.168.5.240:42        192.168.5.189:61374   ESTABLISHED
TCP    192.168.5.240:53        SVRPRINC00:0          LISTENING
TCP    192.168.5.240:135       192.168.5.189:65235   ESTABLISHED
TCP    192.168.5.240:135       SVRPRINC00:63145      ESTABLISHED
TCP    192.168.5.240:139       SVRPRINC00:0          LISTENING
TCP    192.168.5.240:389       SVRPRINC00:62952      ESTABLISHED
TCP    192.168.5.240:389       SVRPRINC00:62963      ESTABLISHED
TCP    192.168.5.240:389       SVRPRINC00:63495      ESTABLISHED
TCP    192.168.5.240:49157     192.168.5.189:50442   ESTABLISHED
TCP    192.168.5.240:49178     SVRPRINC00:63146      ESTABLISHED
TCP    192.168.5.240:62952     SVRPRINC00:ldap       ESTABLISHED
TCP    192.168.5.240:62963     SVRPRINC00:ldap       ESTABLISHED
TCP    192.168.5.240:63145     SVRPRINC00:epmap      ESTABLISHED
TCP    192.168.5.240:63146     SVRPRINC00:49178      ESTABLISHED
TCP    192.168.5.240:63495     SVRPRINC00:ldap       ESTABLISHED
TCP    [::]:80                 SVRPRINC00:0          LISTENING
TCP    [::]:88                 SVRPRINC00:0          LISTENING
TCP    [::]:135                SVRPRINC00:0          LISTENING
TCP    [::]:443                SVRPRINC00:0          LISTENING
TCP    [::]:445                SVRPRINC00:0          LISTENING
TCP    [::]:464                SVRPRINC00:0          LISTENING
. . . . . . . . . . . . . .
TCP    [::]:49205              SVRPRINC00:0          LISTENING
TCP    [::]:49207              SVRPRINC00:0          LISTENING
TCP    [::]:49222              SVRPRINC00:0          LISTENING
TCP    [::]:63163              SVRPRINC00:0          LISTENING
TCP    [::1]:53                SVRPRINC00:0          LISTENING
TCP    [::1]:135               SVRPRINC00:49199      ESTABLISHED
TCP    [::1]:135               SVRPRINC00:49208      ESTABLISHED
TCP    [::1]:49157             SVRPRINC00:49216      ESTABLISHED
TCP    [::1]:49157             SVRPRINC00:49439      ESTABLISHED
TCP    [::1]:49157             SVRPRINC00:63077      ESTABLISHED
TCP    [::1]:49198             SVRPRINC00:49209      ESTABLISHED
TCP    [::1]:49198             SVRPRINC00:49210      ESTABLISHED
TCP    [::1]:49199             SVRPRINC00:epmap      ESTABLISHED
TCP    [::1]:49208             SVRPRINC00:epmap      ESTABLISHED
TCP    [::1]:49209             SVRPRINC00:49198      ESTABLISHED
TCP    [::1]:49210             SVRPRINC00:49198      ESTABLISHED
TCP    [::1]:49216             SVRPRINC00:49157      ESTABLISHED
TCP    [::1]:49439             SVRPRINC00:49157      ESTABLISHED
TCP    [::1]:63077             SVRPRINC00:49157      ESTABLISHED
TCP    [::1]:63613             SVRPRINC00:epmap      TIME_WAIT
UDP    0.0.0.0:42              *:*
UDP    0.0.0.0:123             *:*
UDP    0.0.0.0:389             *:*
UDP    0.0.0.0:3389            *:*
UDP    0.0.0.0:5355            *:*
UDP    0.0.0.0:61212           *:*
..............................
UDP    0.0.0.0:63713           *:*
UDP    127.0.0.1:53            *:*
UDP    127.0.0.1:51957         *:*
UDP    127.0.0.1:51958         *:*
UDP    127.0.0.1:53436         *:*
UDP    127.0.0.1:53437         *:*
UDP    127.0.0.1:54379         *:*
UDP    127.0.0.1:54937         *:*
UDP    127.0.0.1:57301         *:*
UDP    127.0.0.1:58177         *:*
UDP    127.0.0.1:58344         *:*
UDP    127.0.0.1:61209         *:*
UDP    127.0.0.1:61210         *:*
UDP    127.0.0.1:63714         *:*
UDP    127.0.0.1:63715         *:*
UDP    192.168.5.240:53        *:*
UDP    192.168.5.240:67        *:*
UDP    192.168.5.240:68        *:*
UDP    192.168.5.240:88        *:*
UDP    192.168.5.240:137       *:*
UDP    192.168.5.240:138       *:*
UDP    192.168.5.240:464       *:*
UDP    192.168.5.240:2535      *:*
UDP    [::]:123                *:*
UDP    [::]:3389               *:*
UDP    [::]:61213              *:*
UDP    [::1]:53                *:*
UDP    [::1]:61211             *:*
```

> **Ports TCP / IP**
> There are thousands of ports (16-bit encoded, ie it has possibilities 65536).
> IANA (Internet Assigned Numbers Authority [Internet Assigned Numbers]) developed a standard application to help with network configurations.
> - Ports 0 to 1023 are "known ports" or reserved. Generally speaking, they are reserved for system processes (daemons) or programs executed by privileged users. Without em-ever, a network administrator can connect ports services of their choice.
> - Ports 1024 to 49151 are "registered ports".
> - The ports 49152 to 65535 are "dynamic and / or deprives-two ports".

a) Displays the executable involved in creating each connection or listening port and listening ports.

```
C:\Windows\system32>netstat -b

Active Connections

  Proto  Local Address           Foreign Address         State
  TCP    127.0.0.1:389           SVRPRINC00:49163        ESTABLISHED  [lsass.exe]
  TCP    127.0.0.1:389           SVRPRINC00:49165        ESTABLISHED  [lsass.exe]
  TCP    127.0.0.1:389           SVRPRINC00:62961        ESTABLISHED  [lsass.exe]
  TCP    127.0.0.1:49163         SVRPRINC00:ldap         ESTABLISHED  [ismserv.exe]
  TCP    127.0.0.1:49165         SVRPRINC00:ldap         ESTABLISHED  [ismserv.exe]
  TCP    127.0.0.1:62961         SVRPRINC00:ldap         ESTABLISHED  [dns.exe]
  TCP    192.168.5.240:42        192.168.5.189:61374     ESTABLISHED  [wins.exe]
  TCP    192.168.5.240:135       192.168.5.189:65235     ESTABLISHED   RpcSs [svchost.exe]
  TCP    192.168.5.240:135       SVRPRINC00:63145        ESTABLISHED   RpcSs [svchost.exe]
  TCP    192.168.5.240:389       SVRPRINC00:62952        ESTABLISHED  [lsass.exe]
  TCP    192.168.5.240:389       SVRPRINC00:62963        ESTABLISHED  [lsass.exe]
```

```
    TCP    192.168.5.240:49157    192.168.5.189:50442      ESTABLISHED  [lsass.exe]
    TCP    192.168.5.240:49178    SVRPRINC00:63146         ESTABLISHED  [wins.exe]
    TCP    192.168.5.240:62952    SVRPRINC00:ldap          ESTABLISHED  [DFSRs.exe]
    TCP    192.168.5.240:62963    SVRPRINC00:ldap          ESTABLISHED  [DFSRs.exe]
    TCP    192.168.5.240:63145    SVRPRINC00:epmap         ESTABLISHED  [mmc.exe]
    TCP    192.168.5.240:63146    SVRPRINC00:49178         ESTABLISHED  [mmc.exe]
    TCP    192.168.5.240:63702    192.168.5.140:nameserver SYN_SENT     [wins.exe]
    TCP    [::1]:135              SVRPRINC00:49199         ESTABLISHED  RpcSs [svchost.exe]
    TCP    [::1]:135              SVRPRINC00:49208         ESTABLISHED  RpcSs [svchost.exe]
    TCP    [::1]:49157            SVRPRINC00:49216         ESTABLISHED  [lsass.exe]
    TCP    [::1]:49157            SVRPRINC00:49439         ESTABLISHED  [lsass.exe]
    TCP    [::1]:49157            SVRPRINC00:63077         ESTABLISHED  [lsass.exe]
    TCP    [::1]:49198            SVRPRINC00:49209         ESTABLISHED  [tssdis.exe]
    TCP    [::1]:49198            SVRPRINC00:49210         ESTABLISHED  [tssdis.exe]
    TCP    [::1]:49199            SVRPRINC00:epmap         ESTABLISHED  [tssdis.exe]
    TCP    [::1]:49208            SVRPRINC00:epmap         ESTABLISHED  TermService [svchost.exe]
    TCP    [::1]:49209            SVRPRINC00:49198         ESTABLISHED  TermService [svchost.exe]
    TCP    [::1]:49210            SVRPRINC00:49198         ESTABLISHED  TermService [svchost.exe]
    TCP    [::1]:49216            SVRPRINC00:49157         ESTABLISHED  [DFSRs.exe]
    TCP    [::1]:49439            SVRPRINC00:49157         ESTABLISHED  [lsass.exe]
    TCP    [::1]:63077            SVRPRINC00:49157         ESTABLISHED  [Microsoft.ActiveDirectory.WebServices.exe]
    TCP    [::1]:63705            SVRPRINC00:epmap         TIME_WAIT
```

b) Shows the statistic of the Ethernet connection.

```
C:\Windows\system32>NETSTAT -E
Interface Statistics

                           Received            Sent

Bytes                      14978148            20557224
Unicast packets               34604               29180
Non-unicast packets           59032                6828
Discards                          0                   0
Errors                            0                   0
Unknown protocols                 0
```

c) Show fully qualified domain names (FQDN "fully qualified domain name") to external addresses.

```
C:\Windows\system32>NETSTAT -F

Active Connections

  Proto  Local Address           Foreign Address          State
  TCP    127.0.0.1:389           SVRPRINC00.bspWeb.local:49163   ESTABLISHED
  TCP    127.0.0.1:389           SVRPRINC00.bspWeb.local:49165   ESTABLISHED
  TCP    127.0.0.1:389           SVRPRINC00.bspWeb.local:62961   ESTABLISHED
  TCP    127.0.0.1:49163         SVRPRINC00.bspWeb.local:ldap    ESTABLISHED
  TCP    127.0.0.1:49165         SVRPRINC00.bspWeb.local:ldap    ESTABLISHED
  TCP    127.0.0.1:62961         SVRPRINC00.bspWeb.local:ldap    ESTABLISHED
  TCP    192.168.5.240:42        192.168.5.189:61374             ESTABLISHED
  TCP    192.168.5.240:135       192.168.5.189:65235             ESTABLISHED
  TCP    192.168.5.240:135       SVRPRINC00.bspWeb.local:63145   ESTABLISHED
  TCP    192.168.5.240:389       SVRPRINC00.bspWeb.local:62952   ESTABLISHED
  TCP    192.168.5.240:389       SVRPRINC00.bspWeb.local:62963   ESTABLISHED
  TCP    192.168.5.240:49157     192.168.5.189:50442             ESTABLISHED
  TCP    192.168.5.240:49178     SVRPRINC00.bspWeb.local:63146   ESTABLISHED
  TCP    192.168.5.240:62952     SVRPRINC00.bspWeb.local:ldap    ESTABLISHED
  TCP    192.168.5.240:62963     SVRPRINC00.bspWeb.local:ldap    ESTABLISHED
  TCP    192.168.5.240:63145     SVRPRINC00.bspWeb.local:epmap   ESTABLISHED
  TCP    192.168.5.240:63146     SVRPRINC00.bspWeb.local:49178   ESTABLISHED
  TCP    192.168.5.240:63727     192.168.5.189:epmap             SYN_SENT
  TCP    [::1]:135               SVRPRINC00.bspWeb.local:49199   ESTABLISHED
  TCP    [::1]:135               SVRPRINC00.bspWeb.local:49208   ESTABLISHED
  . . . . . . . . . .
  TCP    [::1]:49157             SVRPRINC00.bspWeb.local:49216   ESTABLISHED
  TCP    [::1]:49157             SVRPRINC00.bspWeb.local:49439   ESTABLISHED
  TCP    [::1]:49157             SVRPRINC00.bspWeb.local:63077   ESTABLISHED
  TCP    [::1]:49198             SVRPRINC00.bspWeb.local:49209   ESTABLISHED
  TCP    [::1]:49198             SVRPRINC00.bspWeb.local:49210   ESTABLISHED
  TCP    [::1]:49199             SVRPRINC00.bspWeb.local:epmap   ESTABLISHED
  TCP    [::1]:49208             SVRPRINC00.bspWeb.local:epmap   ESTABLISHED
  TCP    [::1]:49209             SVRPRINC00.bspWeb.local:49198   ESTABLISHED
  TCP    [::1]:49210             SVRPRINC00.bspWeb.local:49198   ESTABLISHED
  TCP    [::1]:49216             SVRPRINC00.bspWeb.local:49157   ESTABLISHED
  TCP    [::1]:49439             SVRPRINC00.bspWeb.local:49157   ESTABLISHED
  TCP    [::1]:63077             SVRPRINC00.bspWeb.local:49157   ESTABLISHED
  TCP    [::1]:63721             SVRPRINC00.bspWeb.local:5985    TIME_WAIT
  TCP    [::1]:63722             SVRPRINC00.bspWeb.local:5985    TIME_WAIT
  TCP    [::1]:63723             SVRPRINC00.bspWeb.local:47001   TIME_WAIT
  TCP    [::1]:63724             SVRPRINC00.bspWeb.local:5985    TIME_WAIT
```

d) Show protocol addresses and ports and their status proprietary process identification (PID) associated with each connection.

```
C:\Windows\system32>NETSTAT -O

Active Connections

  Proto  Local Address          Foreign Address        State           PID
  TCP    127.0.0.1:389          SVRPRINC00:49163       ESTABLISHED     472
  TCP    127.0.0.1:389          SVRPRINC00:49165       ESTABLISHED     472
  TCP    127.0.0.1:389          SVRPRINC00:62961       ESTABLISHED     472
  TCP    127.0.0.1:49163        SVRPRINC00:ldap        ESTABLISHED     1440
  TCP    127.0.0.1:49165        SVRPRINC00:ldap        ESTABLISHED     1440
  TCP    127.0.0.1:62961        SVRPRINC00:ldap        ESTABLISHED     1416
  TCP    192.168.5.240:42       192.168.5.189:61374    ESTABLISHED     1772
  TCP    192.168.5.240:135      192.168.5.189:65235    ESTABLISHED     644
```

```
TCP    192.168.5.240:135        SVRPRINC00:63145         ESTABLISHED    644
TCP    192.168.5.240:389        SVRPRINC00:62952         ESTABLISHED    472
TCP    192.168.5.240:389        SVRPRINC00:62963         ESTABLISHED    472
TCP    192.168.5.240:49157      192.168.5.189:50442      ESTABLISHED    472
TCP    192.168.5.240:49178      SVRPRINC00:63146         ESTABLISHED    1772
TCP    192.168.5.240:62952      SVRPRINC00:ldap          ESTABLISHED    1312
TCP    192.168.5.240:62963      SVRPRINC00:ldap          ESTABLISHED    1312
TCP    192.168.5.240:63145      SVRPRINC00:epmap         ESTABLISHED    676
TCP    192.168.5.240:63146      SVRPRINC00:49178         ESTABLISHED    676
TCP    [::1]:135                SVRPRINC00:49199         ESTABLISHED    644
TCP    [::1]:135                SVRPRINC00:49208         ESTABLISHED    644
  .       .                        .                       .            .
TCP    [::1]:49157              SVRPRINC00:49216         ESTABLISHED    472
TCP    [::1]:49157              SVRPRINC00:49439         ESTABLISHED    472
TCP    [::1]:49157              SVRPRINC00:63077         ESTABLISHED    472
TCP    [::1]:49198              SVRPRINC00:49209         ESTABLISHED    2340
TCP    [::1]:49198              SVRPRINC00:49210         ESTABLISHED    2340
TCP    [::1]:49199              SVRPRINC00:epmap         ESTABLISHED    2340
TCP    [::1]:49208              SVRPRINC00:epmap         ESTABLISHED    2820
TCP    [::1]:49209              SVRPRINC00:49198         ESTABLISHED    2820
TCP    [::1]:49210              SVRPRINC00:49198         ESTABLISHED    2820
TCP    [::1]:49216              SVRPRINC00:49157         ESTABLISHED    1312
TCP    [::1]:49439              SVRPRINC00:49157         ESTABLISHED    472
TCP    [::1]:63077              SVRPRINC00:49157         ESTABLISHED    1224
```

e) Displays the name of PID process associated with the fully qualified domain name (FQDN).

```
C:\Windows\system32>netstat -fo

Active Connections

  Proto  Local Address          Foreign Address                    State          PID
  TCP    127.0.0.1:389          SVRPRINC00.bspWeb.local:49163      ESTABLISHED    472
  TCP    127.0.0.1:389          SVRPRINC00.bspWeb.local:49165      ESTABLISHED    472
  TCP    127.0.0.1:389          SVRPRINC00.bspWeb.local:62961      ESTABLISHED    472
  TCP    127.0.0.1:49163        SVRPRINC00.bspWeb.local:ldap       ESTABLISHED    1440
  TCP    127.0.0.1:49165        SVRPRINC00.bspWeb.local:ldap       ESTABLISHED    1440
  TCP    127.0.0.1:62961        SVRPRINC00.bspWeb.local:ldap       ESTABLISHED    1416
  TCP    192.168.5.240:42       192.168.5.189:61374                ESTABLISHED    1772
  TCP    192.168.5.240:135      192.168.5.189:65235                ESTABLISHED    644
  TCP    192.168.5.240:135      SVRPRINC00.bspWeb.local:63145      ESTABLISHED    644
  TCP    192.168.5.240:389      SVRPRINC00.bspWeb.local:62952      ESTABLISHED    472
  TCP    192.168.5.240:389      SVRPRINC00.bspWeb.local:62963      ESTABLISHED    472
  TCP    192.168.5.240:49157    192.168.5.189:50442                ESTABLISHED    472
  TCP    192.168.5.240:49178    SVRPRINC00.bspWeb.local:63146      ESTABLISHED    1772
  TCP    192.168.5.240:62952    SVRPRINC00.bspWeb.local:ldap       ESTABLISHED    1312
  TCP    192.168.5.240:62963    SVRPRINC00.bspWeb.local:ldap       ESTABLISHED    1312
  TCP    192.168.5.240:63145    SVRPRINC00.bspWeb.local:epmap      ESTABLISHED    676
  TCP    192.168.5.240:63146    SVRPRINC00.bspWeb.local:49178      ESTABLISHED    676
  TCP    [::1]:135              SVRPRINC00.bspWeb.local:49199      ESTABLISHED    644
  TCP    [::1]:135              SVRPRINC00.bspWeb.local:49208      ESTABLISHED    644
  TCP    [::1]:49157            SVRPRINC00.bspWeb.local:49216      ESTABLISHED    472
  TCP    [::1]:49157            SVRPRINC00.bspWeb.local:49439      ESTABLISHED    472
  TCP    [::1]:49157            SVRPRINC00.bspWeb.local:63077      ESTABLISHED    472
  TCP    [::1]:49198            SVRPRINC00.bspWeb.local:49209      ESTABLISHED    2340
  TCP    [::1]:49198            SVRPRINC00.bspWeb.local:49210      ESTABLISHED    2340
  TCP    [::1]:49199            SVRPRINC00.bspWeb.local:epmap      ESTABLISHED    2340
  TCP    [::1]:49208            SVRPRINC00.bspWeb.local:epmap      ESTABLISHED    2820
  TCP    [::1]:49209            SVRPRINC00.bspWeb.local:49198      ESTABLISHED    2820
  TCP    [::1]:49210            SVRPRINC00.bspWeb.local:49198      ESTABLISHED    2820
  TCP    [::1]:49216            SVRPRINC00.bspWeb.local:49157      ESTABLISHED    1312
  TCP    [::1]:49439            SVRPRINC00.bspWeb.local:49157      ESTABLISHED    472
  TCP    [::1]:63077            SVRPRINC00.bspWeb.local:49157      ESTABLISHED    1224
```

f) Displays the name of PID process associated with the fully qualified domain name (FQDN) and displays the connections and listening ports, the number of addresses and listening ports.

```
C:\Windows\system32>NETSTAT -FANO

Active Connections

  Proto  Local Address          Foreign Address        State           PID
  TCP    0.0.0.0:42             0.0.0.0:0              LISTENING       1772
  TCP    0.0.0.0:80             0.0.0.0:0              LISTENING       4
  TCP    0.0.0.0:88             0.0.0.0:0              LISTENING       472
  TCP    0.0.0.0:135            0.0.0.0:0              LISTENING       644
  TCP    0.0.0.0:389            0.0.0.0:0              LISTENING       472
  TCP    0.0.0.0:443            0.0.0.0:0              LISTENING       4
  TCP    0.0.0.0:445            0.0.0.0:0              LISTENING       4
  TCP    0.0.0.0:464            0.0.0.0:0              LISTENING       472
  TCP    0.0.0.0:593            0.0.0.0:0              LISTENING       644
  TCP    0.0.0.0:636            0.0.0.0:0              LISTENING       472
  TCP    0.0.0.0:3268           0.0.0.0:0              LISTENING       472
  TCP    0.0.0.0:3269           0.0.0.0:0              LISTENING       472
  TCP    0.0.0.0:3389           0.0.0.0:0              LISTENING       2820
  TCP    0.0.0.0:5504           0.0.0.0:0              LISTENING       2316
  TCP    0.0.0.0:5985           0.0.0.0:0              LISTENING       4
  TCP    0.0.0.0:9389           0.0.0.0:0              LISTENING       1224
  TCP    0.0.0.0:47001          0.0.0.0:0              LISTENING       4
  TCP    0.0.0.0:49152          0.0.0.0:0              LISTENING       376
  TCP    0.0.0.0:49153          0.0.0.0:0              LISTENING       840
  TCP    0.0.0.0:49154          0.0.0.0:0              LISTENING       472
  TCP    0.0.0.0:49155          0.0.0.0:0              LISTENING       872
  TCP    0.0.0.0:49157          0.0.0.0:0              LISTENING       472
  TCP    0.0.0.0:49158          0.0.0.0:0              LISTENING       472
  TCP    0.0.0.0:49159          0.0.0.0:0              LISTENING       1196
  TCP    0.0.0.0:49173          0.0.0.0:0              LISTENING       1416
```

```
TCP    0.0.0.0:49178          0.0.0.0:0              LISTENING      1772
TCP    0.0.0.0:49179          0.0.0.0:0              LISTENING      1396
TCP    0.0.0.0:49198          0.0.0.0:0              LISTENING      2340
TCP    0.0.0.0:49205          0.0.0.0:0              LISTENING      464
TCP    0.0.0.0:49207          0.0.0.0:0              LISTENING      2820
TCP    0.0.0.0:49222          0.0.0.0:0              LISTENING      1312
TCP    0.0.0.0:63163          0.0.0.0:0              LISTENING      524
TCP    127.0.0.1:53           0.0.0.0:0              LISTENING      1416
TCP    127.0.0.1:389          127.0.0.1:49163        ESTABLISHED    472
TCP    127.0.0.1:389          127.0.0.1:49165        ESTABLISHED    472
TCP    127.0.0.1:389          127.0.0.1:62961        ESTABLISHED    472
TCP    127.0.0.1:49163        127.0.0.1:389          ESTABLISHED    1440
TCP    127.0.0.1:49165        127.0.0.1:389          ESTABLISHED    1440
TCP    127.0.0.1:62961        127.0.0.1:389          ESTABLISHED    1416
TCP    192.168.5.240:42       192.168.5.189:61374    ESTABLISHED    1772
TCP    192.168.5.240:53       0.0.0.0:0              LISTENING      1416
TCP    192.168.5.240:135      192.168.5.189:65235    ESTABLISHED    644
TCP    192.168.5.240:135      192.168.5.240:63145    ESTABLISHED    644
TCP    192.168.5.240:139      0.0.0.0:0              LISTENING      4
TCP    192.168.5.240:389      192.168.5.240:62952    ESTABLISHED    472
TCP    192.168.5.240:389      192.168.5.240:62963    ESTABLISHED    472
TCP    192.168.5.240:49157    192.168.5.189:50442    ESTABLISHED    472
TCP    192.168.5.240:49178    192.168.5.240:63146    ESTABLISHED    1772
TCP    192.168.5.240:62952    192.168.5.240:389      ESTABLISHED    1312
TCP    192.168.5.240:62963    192.168.5.240:389      ESTABLISHED    1312
TCP    192.168.5.240:63145    192.168.5.240:135      ESTABLISHED    676
TCP    192.168.5.240:63146    192.168.5.240:49178    ESTABLISHED    676
TCP    192.168.5.240:63748    192.168.5.189:42       SYN_SENT       1772
TCP    [::]:80                [::]:0                 LISTENING      4
TCP    [::]:88                [::]:0                 LISTENING      472
TCP    [::]:135               [::]:0                 LISTENING      644
TCP    [::]:443               [::]:0                 LISTENING      4
TCP    [::]:445               [::]:0                 LISTENING      4
TCP    [::]:464               [::]:0                 LISTENING      472
TCP    [::]:593               [::]:0                 LISTENING      644
. . . . . . . . . .
UDP    0.0.0.0:63713          *:*                                   1416
UDP    127.0.0.1:53           *:*                                   1416
UDP    127.0.0.1:51957        *:*                                   1224
UDP    127.0.0.1:51958        *:*                                   2564
UDP    127.0.0.1:53436        *:*                                   1312
UDP    127.0.0.1:53437        *:*                                   3060
UDP    127.0.0.1:54379        *:*                                   872
UDP    127.0.0.1:54937        *:*                                   1772
UDP    127.0.0.1:57301        *:*                                   472
UDP    127.0.0.1:58177        *:*                                   160
UDP    127.0.0.1:58344        *:*                                   2424
UDP    127.0.0.1:61209        *:*                                   1396
UDP    127.0.0.1:61210        *:*                                   1440
UDP    127.0.0.1:63714        *:*                                   1416
UDP    127.0.0.1:63715        *:*                                   984
UDP    192.168.5.240:53       *:*                                   1416
UDP    192.168.5.240:67       *:*                                   1396
UDP    192.168.5.240:68       *:*                                   1396
UDP    192.168.5.240:88       *:*                                   472
UDP    192.168.5.240:137      *:*                                   4
UDP    192.168.5.240:138      *:*                                   4
UDP    192.168.5.240:464      *:*                                   472
UDP    192.168.5.240:2535     *:*                                   1396
UDP    [::]:123               *:*                                   920
UDP    [::]:3389              *:*                                   2820
UDP    [::]:61213             *:*                                   1416
UDP    [::1]:53               *:*                                   1416
UDP    [::1]:61211            *:*                                   1416
```

g) Sample connections, listeners and shared NetworkDirect ends. If none exists will not show results.
```
C:\Windows\system32>NETSTAT    -X

Active NetworkDirect Connections, Listeners, SharedEndpoints

   Mode    IfIndex  Type       Local Address          Foreign Address           PID
```

h) Displays templates TCP connection for all connections. The template (DATECENTER).
```
C:\Windows\system32>NETSTAT    -Y

Active Connections

   Proto  Local Address          Foreign Address        State           Template

   TCP    127.0.0.1:389          SVRPRINC00:49161       ESTABLISHED     Datacenter
   TCP    127.0.0.1:389          SVRPRINC00:49163       ESTABLISHED     Datacenter
   TCP    127.0.0.1:389          SVRPRINC00:60471       ESTABLISHED     Datacenter
   TCP    127.0.0.1:49161        SVRPRINC00:ldap        ESTABLISHED     Datacenter
   TCP    127.0.0.1:49163        SVRPRINC00:ldap        ESTABLISHED     Datacenter
   TCP    192.168.2.89:49284     www:http               ESTABLISHED     Internet
   TCP    192.168.2.89:49288     www:http               ESTABLISHED     Internet
   TCP    192.168.2.89:49357     ekb-db:http            ESTABLISHED     Internet
   TCP    192.168.2.89:49363     mad06s09-in-f130:https CLOSE_WAIT      Internet
   TCP    192.168.2.89:49372     92.42.227.57:http      ESTABLISHED     Internet
   TCP    192.168.2.89:49373     92.42.227.57:http      ESTABLISHED     Internet
   TCP    127.0.0.1:60471        SVRPRINC00:ldap        ESTABLISHED     Datacenter
```

i) Sample connections, listeners and shared NetworkDirect ends. But there will not appear any results and displays the template TCP connection for all connections. Can not be combined with other options, as seen not give any results.

```
C:\Windows\system32>netstat -xy

Active NetworkDirect Connections, Listeners, SharedEndpoints

   Mode   IfIndex Type           Local Address            Foreign Address            PID
```

j) Display connections for the protocol specified by protocol.
```
C:\Windows\system32>NETSTAT -P UDP

Active Connections

  Proto  Local Address          Foreign Address        State

C:\Windows\system32>NETSTAT -p UDPv6

Active Connections

  Proto  Local Address          Foreign Address        State

C:\Windows\system32>NETSTAT -p TCP


Active Connections

  Proto  Local Address          Foreign Address        State
  TCP    127.0.0.1:389          SVRPRINC00:49161       ESTABLISHED
  TCP    127.0.0.1:389          SVRPRINC00:49163       ESTABLISHED
  TCP    127.0.0.1:389          SVRPRINC00:60471       ESTABLISHED
  TCP    127.0.0.1:49161        SVRPRINC00:ldap        ESTABLISHED
  TCP    127.0.0.1:49163        SVRPRINC00:ldap        ESTABLISHED
  TCP    127.0.0.1:60471        SVRPRINC00:ldap        ESTABLISHED
  TCP    192.168.2.89:49284     www:http               ESTABLISHED
  TCP    192.168.2.89:49288     www:http               ESTABLISHED
  TCP    192.168.2.89:49357     ekb-db:http            ESTABLISHED
  TCP    192.168.2.89:49363     mad06s09-in-f130:https CLOSE_WAIT
  TCP    192.168.2.89:49372     92.42.227.57:http      ESTABLISHED
  TCP    192.168.2.89:49373     92.42.227.57:http      ESTABLISHED
```

## STEP 2: Run from a Windows 10 Professional. NBTSTAT

a) Make a list of table names of the remote computers by name.
```
C:\Users\aprendiz>nbtstat -a www.google.es

vEthernet (My network card i7):
Node IP Address: [192.168.1.99] field ID:. []

    Host not found.

Wireless network connection:
Node IP Address: [0.0.0.0] ID field. []

    Host not found.

Local Area Connection * 2:
Node IP Address: [0.0.0.0] ID field. []

    Host not found.

Local connection * 4 area:
Node IP Address: [0.0.0.0] ID field. []

    Host not found.

2 red Bluetooth connection:
Node IP Address: [0.0.0.0] ID field. []

    Host not found.
```

a.1) Make a list of table names of the remote computers by name.
```
C:\Users\aprendiz>NBTSTAT -a google

vEthernet (My network card i7):
Node IP Address: [192.168.1.99] field ID:. []

    Host not found.

Wireless network connection:
Node IP Address: [0.0.0.0] ID field. []

    Host not found.

Local Area Connection * 2:
Node IP Address: [0.0.0.0] ID field. []

    Host not found.

Local connection * 4 area:
Node IP Address: [0.0.0.0] ID field. []

    Host not found.

2 red Bluetooth connection:
```

```
        Node IP Address: [0.0.0.0] ID field. []

            Host not found.
```
a.2) Make a list of table names for remote computers by IP address.
```
        C:\Users\aprendiz>NBTSTAT -a   8.8.8.8

        vEthernet (My network card i7):
        Node IP Address: [192.168.1.99] field ID:. []

            Host not found.

        Wireless network connection:
        Node IP Address: [0.0.0.0] ID field. []

            Host not found.

        Local Area Connection * 2:
        Node IP Address: [0.0.0.0] ID field. []

            Host not found.

        Local connection * 4 area:
        Node IP Address: [0.0.0.0] ID field. []

            Host not found.

        2 red Bluetooth connection:
        Node IP Address: [0.0.0.0] ID field. []

            Host not found.
```
b) Performing displays a list of table names of the remote computers as their IP addresses, established.
```
        C:\Users\aprendiz>NBTSTAT -A 192.168.1.99

        vEthernet (My network card i7):
        Node IP Address: [192.168.1.99] field ID:. []

            Name table NetBIOS remote computers

               Name Type Status
           ---------------------------------------------
           I7-PC <20> Unique Registered
           I7-PC <00> Unique Registered
           WORKGROUP <00> Group Joined
           WORKGROUP <1E> Group Joined
           WORKGROUP <1D> Unique Registered
              ☺●__MSBROWSE__●<01> Join Group

           MAC address = 5C-F9-DD-40-96-17

        Wireless network connection:
        Node IP Address: [0.0.0.0] ID field. []

            Host not found.

        Local Area Connection * 2:
        Node IP Address: [0.0.0.0] ID field. []

            Host not found.

        Local Area Connection * 2:
        Node IP Address: [0.0.0.0] ID field. []

            Host not found.

        Local connection * 4 area:
        Node IP Address: [0.0.0.0] ID field. []

            Host not found.

        2 red Bluetooth connection:
        Node IP Address: [0.0.0.0] ID field. []

            Host not found.
```
c) Conducting displays a list of table names of the remote computers as their IP addresses, stable-cide. A list of names [computer] NBT remote cache and IP addresses ago.
```
        C:\Users\aprendiz>nbtstat -A 192.168.1.99 -c
        vEthernet (My network card i7):
        Node IP Address: [192.168.1.99] field ID:. []

                   NetBIOS Remote Cache Table

              Host Name Type Dir Life [s]
           ---------------------------------------------  ----------
              I7-PC <20> Unique 192.168.1.99 542

        Wireless network connection:
        Node IP Address: [0.0.0.0] ID field. []

            No names in cache
```

```
Local Area Connection * 2:
Node IP Address: [0.0.0.0] ID field. []

    No names in cache

Local connection * 4 area:
Node IP Address: [0.0.0.0] ID field. []

    No names in cache

2 red Bluetooth connection:
Node IP Address: [0.0.0.0] ID field. []

No names in cache
```

d) Make a list of the remote cache name.

```
C:\Users\aprendiz>nbtstat -c

vEthernet (My network card i7):
Node IP Address: [192.168.1.99] field ID:. []

                NetBIOS Remote Cache Table

    Host Name Type Dir Life [s]
    ---------------------------------------------- ----------
    I7-PC <20> Unique 192.168.1.99 528

Wireless network connection:
Node IP Address: [0.0.0.0] ID field. []

    No names in cache

Local Area Connection * 2:
Node IP Address: [0.0.0.0] ID field. []

    No names in cache

Local connection * 4 area:
Node IP Address: [0.0.0.0] ID field. []

    No names in cache

2 red Bluetooth connection:
Node IP Address: [0.0.0.0] ID field. []

    No names in cache
```

e) A list of local NetBIOS names ago.

```
C:\Users\aprendiz>nbtstat -n

vEthernet (My network card i7):
Node IP Address: [192.168.1.99] field ID:. []

                NetBIOS local name table

    Name Type Status
    ---------------------------------------------
    I7-PC        <20> Unique Registered
    I7-PC        <00> Unique Registered
    WORKGROUP    <00> Group Joined
    WORKGROUP    <1E> Group Joined
    WORKGROUP    <1D> Unique Registered
    ☺●__MSBROWSE__●<01> Join Group

Wireless network connection:
Node IP Address: [0.0.0.0] ID field. []

    No names in cache

Local Area Connection * 2:
Node IP Address: [0.0.0.0] ID field. []

    No names in cache

Local connection * 4 area:
Node IP Address: [0.0.0.0] ID field. []

    No names in cache

2 red Bluetooth connection:
Node IP Address: [0.0.0.0] ID field. []

    No names in cache
```

f) List names resolved by broadcast and via WINS.

```
C:\Users\aprendiz>NBTSTAT -r

    Statistics resolution and NetBIOS name registration
    ----------------------------------------------------

    Resolved by broadcast = 0,
    Resolved by the nameserver = 0

    Registered diffusion = 134
    Registered by the nameserver = 0
```

g) Purge and reload the name table of the remote cache.
```
C:\Users\aprendiz>nbtstat -R
    You can not purge the table NBT Remote Cache.
    You can not purge the table NBT Remote Cache.
    You can not purge the table NBT Remote Cache.
    You can not purge the table NBT Remote Cache.
    You can not purge the table NBT Remote Cache.
```
h) Lists sessions table converting destination IP addresses to NetBIOS computer names.
```
C:\Users\aprendiz>nbtstat -s
vEthernet (My network card i7):
Node IP Address: [192.168.1.99] field ID:. []

    No connection

Wireless network connection:
Node IP Address: [0.0.0.0] ID field. []

    No connection

Local Area Connection * 2:
Node IP Address: [0.0.0.0] ID field. []

    No connection

Local connection * 4 area:
Node IP Address: [0.0.0.0] ID field. []

    No connection

2 red Bluetooth connection:
Node IP Address: [0.0.0.0] ID field. []

    No connection
```

i) Displays or shows the embodiment of a list of the sessions table with the destination addresses of IP.
```
C:\Users\aprendiz>nbtstat -S
vEthernet (My network card i7):
Node IP Address: [192.168.1.99] field ID:. []

    No connection

Wireless network connection:
Node IP Address: [0.0.0.0] ID field. []

    No connection

Local Area Connection * 2:
Node IP Address: [0.0.0.0] ID field. []

    No connection

Local connection * 4 area:
Node IP Address: [0.0.0.0] ID field. []

    No connection

2 red Bluetooth connection:
Node IP Address: [0.0.0.0] ID field. []

    No connection
```

## STEP 3: Run from a Windows 2012 Server domain. NBTSTAT

a) A list of the names [computer] NBT remote cache and IP addresses ago.
```
C:\Windows\SYSVOL\sysvol\bspWeb.local>nbtstat -c

Ethernet:
Node IpAddress: [192.168.1.89] Scope Id: []

    No names in cache
```
b) A list of local NetBIOS names ago.
```
C:\Windows\SYSVOL\sysvol\bspWeb.local>nbtstat -n

Ethernet:
Node IpAddress: [192.168.1.89] Scope Id: []

            NetBIOS Local Name Table

    Name            Type         Status
    ---------------------------------------
    SVRPRINC00   <00>  UNIQUE    Registered
    BSPWEB0      <00>  GROUP     Registered
    BSPWEB0      <1C>  GROUP     Registered
    SVRPRINC00   <20>  UNIQUE    Registered
    BSPWEB0      <1B>  UNIQUE    Registered
```

c) Make a list of names [computer] NBT remote cache and IP addresses.
```
C:\Windows\system32>NBTSTAT -c
```

```
Ethernet:
Node IpAddress: [192.168.1.89] Scope Id: []

    No names in cache
```

d) Lists sessions table converting destination IP addresses to NetBIOS computer names.
```
C:\Windows\system32>NBTSTAT -s

Ethernet:
Dirección IP del nodo: [192.168.2.40] Id. de ámbito : []

    No hay conexiones
```

e) List names resolved by broadcast and via WINS.
```
C:\Windows\system32>NBTSTAT -r

    NetBIOS Names Resolution and Registration Statistics
    ----------------------------------------------------

    Resolved By Broadcast     = 0
    Resolved By Name Server   = 0

    Registered By Broadcast   = 5
    Registered By Name Server = 0
```

f) Displays or shows the embodiment of a list of the sessions table with the destination addresses of IP.
```
C:\Windows\SYSVOL\sysvol\bspWeb.local>nbtstat -S

Ethernet:
Node IpAddress: [192.168.1.89] Scope Id: []

    No Connections
```

g) Clean or purge and reloads the table NBT Remote Cache Name.
```
C:\Windows\SYSVOL\sysvol\bspWeb.local>NBTSTAT -R
    Successful purge and preload of the NBT Remote Cache Name Table.
```

h) To update the NetBIOS names that are registered on this computer.
```
C:\Windows\SYSVOL\sysvol\bspWeb.local>nbtstat -RR
    The NetBIOS names registered by this computer have been refreshed.

C:\Windows\SYSVOL\sysvol\bspWeb.local>nbtstat -RR
    Failed Release and Refresh of Registered names
    Please retry after 2 minutes

C:\Windows\SYSVOL\sysvol\bspWeb.local>nbtstat -Rr

    NetBIOS Names Resolution and Registration Statistics
    ----------------------------------------------------

    Resolved By Broadcast     = 0
    Resolved By Name Server   = 0

    Registered By Broadcast   = 5
    Registered By Name Server = 0

C:\Windows\system32>nbtstat -A 192.168.2.40

C:\Windows\SYSVOL\sysvol\bspWeb.local> nbtstat -RR

    Resolution NetBIOS Names Registration and Statistics
    ---------------------------------------------------- -

    Resolved By Broadcast = 0
    Resolved By Name Server = 0

    Registered By Broadcast = 5
    Registered By Name Server = 0
```

i) Make a list of table names of the remote computers by name.
```
C:\Windows\system32>nbtstat -a 192.168.2.40

Ethernet:
Node IP Address: [192.168.2.40] field ID:. []

    Name table NetBIOS remote computers

       Name          Type     Status
    ---------------------------------------------
    SVR-BSP-00 <20> Unique Registered
    SVR-BSP-00 <00> Unique Registered
    WORKGROUP  <00> Group  Registered

    MAC address = 00-15-5D-01-63-00
Note: card information server
```

j) Check all the information on the card and all components.
```
C:\Windows\SYSVOL\sysvol\bspWeb.local>ipconfig /allcompartments /all

Windows IP Configuration

================================================================================
Network Information for Compartment 1 (ACTIVE)
================================================================================
   Host Name . . . . . . . . . . . . : SVRPRINC00
```

```
        Primary Dns Suffix . . . . . . . : bspWeb.local
        Node Type . . . . . . . . . . . . : Hybrid
        IP Routing Enabled. . . . . . . . : No
        WINS Proxy Enabled. . . . . . . . : No
        DNS Suffix Search List. . . . . . : bspWeb.local

    Ethernet adapter Ethernet:

        Connection-specific DNS Suffix  . :
        Description . . . . . . . . . . . : Microsoft Hyper-V Network Adapter
        Physical Address. . . . . . . . . : 00-15-5D-01-63-06
        DHCP Enabled. . . . . . . . . . . : No
        Autoconfiguration Enabled . . . . : Yes
        IPv4 Address. . . . . . . . . . . : 192.168.1.89(Preferred)
        Subnet Mask . . . . . . . . . . . : 255.255.255.0
        IPv4 Address. . . . . . . . . . . : 192.168.2.89(Preferred)
        Subnet Mask . . . . . . . . . . . : 255.255.255.0
        IPv4 Address. . . . . . . . . . . : 192.168.5.240(Preferred)
        Subnet Mask . . . . . . . . . . . : 255.255.255.0
        Default Gateway . . . . . . . . . : 192.168.2.100
                                            192.168.1.1
        DNS Servers . . . . . . . . . . . : 127.0.0.1
        NetBIOS over Tcpip. . . . . . . . : Enabled

    Tunnel adapter isatap.{0D7F8ED4-1EC1-4BB3-A058-D0C317F404A5}:

        Media State . . . . . . . . . . . : Media disconnected
        Connection-specific DNS Suffix  . :
        Description . . . . . . . . . . . : Microsoft ISATAP Adapter #2
        Physical Address. . . . . . . . . : 00-00-00-00-00-00-00-E0
        DHCP Enabled. . . . . . . . . . . : No
        Autoconfiguration Enabled . . . . : Yes
```

## STEP 4: Check and fix errors cooling and removing NetBIOS cache.

Have joined in a same command line 3 commands that should be run whenever the first run the following runs and then the last.

    NBTSTAT –R & NBTSTAT -r & NBTSTAT –c

a) Executed in Windows 10.

```
C:\Windows\system32>NBTSTAT -R & NBTSTAT -r & NBTSTAT -c
Purging and correct preload table NBT Remote Cache Name.

    Statistics resolution and NetBIOS name registration
    ------------------------------------------------ ------

    Resolved by broadcast = 54
    Resolved by the nameserver = 0

    Registered diffusion = 62
    Registered by the nameserver = 0

    NetBIOS names resolved by broadcast
    ---------------------------------------------
                 ▓ □ 坩 ~ ††††  ▓ □ 坩 ~ ††††
                 ▓ □ 坩 ~ ††††  ▓ □ 坩 ~ ††††
                 ▓ □ 坩 ~ ††††  ▓ □ 坩 ~ ††††
                 ▓ □ 坩 ~ ††††  ▓ □ 坩 ~ ††††
                 ▓ □ 坩 ~ ††††  ▓ □ 坩 ~ ††††
                 ▓ □ 坩 ~ ††††  ▓ □ 坩 ~ ††††
                 ▓ □ 坩 ~ ††††  ▓ □ 坩 ~ ††††
                       ▓ □ 坩 ~ ††††

vEthernet (My network card i7):
Node IP Address: [0.0.0.0] ID field. []

    No names in cache

2 red Bluetooth connection:
Node IP Address: [0.0.0.0] ID field. []

    No names in cache

Wireless network connection:
Node IP Address: [192.168.2.188] ID field. []

    No names in cache

Local Area Connection * 2:
Node IP Address: [0.0.0.0] ID field. []

    No names in cache
```

b) Executed in Windows 2012 Server.

```
C:\Windows\system32>NBTSTAT -R & NTBSTAT -r & NBTSTAT  -c
   Purging and correct preload table NBT Remote Cache Name.
"NTBSTAT" is not recognized as an internal or external command,
program or batch file.

Ethernet:
Node IP Address: [192.168.2.40] field ID:. []

    No names in cache
```

b.1) Second Server.
```
C:\Windows\system32>NBTSTAT -R & NBTSTAT -r & NBTSTAT  -c
    Purging and correct preload table NBT Remote Cache Name.

    Statistics resolution and NetBIOS name registration
    ---------------------------------------------- ------

       Resolved by broadcast = 0,
       Resolved by the nameserver = 0

       Registered diffusion = 5
       Registered by the nameserver = 0

Ethernet:
Node IP Address: [192.168.2.40] field ID:. []

       No names in cache
```

## STEP 5: Indicate route and show their track to the destination. TRACERT

The TRACERT command, the route is shown from the request and all the host that handled the request until it reaches the destination host and returns this communication. The type of IP packets that are sent to it and uses the header field TTL, direntes assigning values to go by calculating the range. We report the latency of each packet is an estimate of the distance that are both ends of the communication.

a) Help.

    TRACERT /?

b) Show route default, no options.
```
C:\Windows\SYSVOL\sysvol\bspWeb.local>tracert  8.8.8.8

Tracing route to google-public-dns-a.google.com [8.8.8.8]
over a maximum of 30 hops:

  1     1 ms    <1 ms    <1 ms  192.168.2.100
  2     7 ms     7 ms     7 ms  10.237.160.1
  3     8 ms     7 ms     6 ms  10.105.240.65
  4    13 ms    15 ms    13 ms  10.254.14.165
  5    16 ms    13 ms    16 ms  10.254.2.141
  6    14 ms    14 ms    15 ms  10.254.3.222
  7    16 ms    14 ms    15 ms  62.42.228.62.static.user.ono.com [62.42.228.62]
  8    15 ms    14 ms    15 ms  72.14.234.231
  9    16 ms    14 ms    16 ms  216.239.48.135
 10    14 ms    15 ms    17 ms  google-public-dns-a.google.com [8.8.8.8]

Trace complete.

C:\Windows\SYSVOL\sysvol\bspWeb.local>tracert  www.google.com

Tracing route to www.google.com [216.58.201.132]
over a maximum of 30 hops:

  1    <1 ms    <1 ms     1 ms  192.168.2.100
  2     9 ms     5 ms     7 ms  10.237.160.1
  3    14 ms     7 ms     7 ms  10.105.240.141
  4    19 ms    16 ms    13 ms  10.254.14.165
  5    14 ms    16 ms    16 ms  10.254.13.241
  6    15 ms    15 ms    14 ms  10.254.10.150
  7    16 ms    16 ms    14 ms  62.42.228.62.static.user.ono.com [62.42.228.62]
  8    16 ms    16 ms    16 ms  72.14.235.18
  9    15 ms    15 ms    14 ms  216.239.40.217
 10    16 ms    15 ms    15 ms  mad06s25-in-f4.1e100.net [216.58.201.132]

Trace complete.
```

c) Do not show the domain name to the destination.
```
C:\Windows\system32>tracert -d www.google.es

Tracing route to www.google.es [74.125.206.94]
over a maximum of 30 hops:
  1     2 ms     2 ms     1 ms  192.168.2.100
  2     8 ms    12 ms     9 ms  10.237.160.1
  3     8 ms     7 ms     8 ms  10.105.240.141
  4    14 ms    16 ms    13 ms  10.105.240.157
  5    26 ms    15 ms    25 ms  10.254.2.145
  6    14 ms    16 ms    20 ms  10.254.10.150
  7    51 ms    47 ms    46 ms  62.42.228.62
  8    23 ms    16 ms    21 ms  72.14.235.20
  9    38 ms    33 ms    32 ms  209.85.245.237
 10    35 ms    38 ms    42 ms  216.239.57.237
 11    92 ms    83 ms    81 ms  216.239.62.155
 12     *        *        *     Request timed out.
 13     *        *        *     Request timed out.
 14     *        *        *     Request timed out.
 15     *        *        *     Request timed out.
 16     *        *        *     Request timed out.
 17     *        *        *     Request timed out.
 18     *        *        *     Request timed out.
 19     *        *        *     Request timed out.
 20   149 ms   133 ms   230 ms  74.125.206.94
```

> **Traceroute:** calculated time further statistical RTT or network latency of these packets, this is an estimate as to the distance to which are the ends of the communication.
>
> **Round-trip delay time (RTT).**
> Used in telecommunications as the time it takes for a packet that is sent from the sender to the receiver and returns this if it reaches. Eg. From 1.0 use HTTP, TCP, in the previous download.

```
Trace complete.

C:\Windows\system32>tracert -d 8.8.8.8

Trace paths 8.8.8.8 on 30 hops maximum.

  1     4 ms     6 ms     3 ms  192.168.2.100
  2   332 ms    22 ms    21 ms  10.237.160.1
  3     7 ms    10 ms     8 ms  10.105.240.65
  4    17 ms    17 ms    28 ms  10.254.14.165
  5    15 ms    35 ms    15 ms  10.254.2.137
  6    18 ms    16 ms    21 ms  10.254.10.150
  7    50 ms    49 ms    48 ms  62.42.228.62
  8    17 ms    21 ms    19 ms  72.14.233.161
  9    17 ms    21 ms    18 ms  216.239.48.109
 10    18 ms    37 ms    15 ms  8.8.8.8

Trace complete.

C:\Windows\system32>tracert  -d www.google.com

Tracing route to www.google.com [74.125.206.99]
over a maximum of 30 hops:

  1     3 ms     3 ms     1 ms  192.168.2.100
  2     7 ms    10 ms     8 ms  10.237.160.1
  3     9 ms     9 ms     8 ms  10.105.240.165
  4    16 ms    28 ms    16 ms  10.254.14.165
  5    22 ms    18 ms    18 ms  10.254.7.57
  6    18 ms    16 ms    14 ms  10.254.3.222
  7    55 ms    50 ms    56 ms  62.42.228.62
  8    16 ms    18 ms    16 ms  72.14.235.18
  9    31 ms    31 ms    32 ms  209.85.246.133
 10    45 ms    36 ms    64 ms  216.239.50.165
 11    35 ms    37 ms    35 ms  216.239.47.177
 12     *        *        *     Request timed out.
 13     *        *        *     Request timed out.
 14     *        *        *     Request timed out.
 15     *        *        *     Request timed out.
 16     *        *        *     Request timed out.
 17     *        *        *     Request timed out.
 18     *        *        *     Request timed out.
 19     *        *        *     Request timed out.
 20    37 ms    35 ms    35 ms  74.125.206.99

Trace complete.
```

d) Set the maximum number of hops.
   TRACERT  -H (valor)

```
C:\Windows\system32>tracert -h 14 8.8.8.8

Tracing route to google-public-dns-a.google.com [8.8.8.8]
over a maximum of 14 hops::

  1     3 ms     6 ms     2 ms  192.168.2.100
  2     8 ms     7 ms     7 ms  10.237.160.1
  3    15 ms   130 ms     7 ms  10.105.240.65
  4    17 ms    15 ms    18 ms  10.254.14.165
  5    16 ms    14 ms    27 ms  10.254.2.137
  6    15 ms    41 ms    29 ms  10.254.10.150
  7    16 ms    17 ms    20 ms  62.42.228.62.static.user.ono.com [62.42.228.62]
  8    15 ms    16 ms    18 ms  72.14.233.161
  9    17 ms    19 ms    16 ms  216.239.48.109
 10    19 ms    18 ms    17 ms  google-public-dns-a.google.com [8.8.8.8]

Trace complete.

C:\Windows\system32>tracert -h 14 www.google.com

Tracing route to www.google.com [74.125.206.99]
over a maximum of 14 hops:

  1     4 ms     3 ms     1 ms  192.168.2.100
  2     8 ms     8 ms     8 ms  10.237.160.1
  3     9 ms     7 ms     7 ms  10.105.240.165
  4    17 ms    25 ms    13 ms  10.254.14.165
  5    31 ms    15 ms    15 ms  10.254.7.57
  6    25 ms    22 ms    20 ms  10.254.3.222
  7    17 ms    18 ms    26 ms  62.42.228.62.static.user.ono.com [62.42.228.62]
  8    19 ms    15 ms    15 ms  72.14.235.18
  9    36 ms    44 ms   102 ms  209.85.246.133
 10    43 ms    63 ms    54 ms  216.239.50.165
 11    40 ms    37 ms    35 ms  216.239.47.177
 12     *        *        *     Request timed out.
 13     *        *        *     Request timed out.
 14     *        *        *     Request timed out.

Trace complete.
```

e) Establish the list of computers for which you must pass.
   e.1) Wrong list.

```
C:\Windows\system32>TRACERT -j 8.8.8.8 8.8.4.4  4.4.4.4

Tracing route to alu7750testscr.xyz1.gblx.mgmt.Level3.net [4.4.4.4]
over a maximum of 30 hops:

  1     *        *        *     Request timed out.
  2     *        *        *     Request timed out.
  . . . . .
 30     *        *        *     Request timed out.

Traza completa.
```

> The maximum number of IP addresses that form the list is 9.

```
C:\Users\aprendiz>TRACERT -j 10.12.0.1 10.29.3.1  10.1.44.1 corp7.microsoft.com

Tracing route to corp7.microsoft.com [92.242.134.28]
over a maximum of 30 hops:

  1     *        *        *     Request timed out.
  2     *        *        *     Request timed out.
  3     *        *        *     Request timed out.
  4     *      ^C

C:\Users\aprendiz>TRACERT -j 192.168.2.100 8.8.8.8 corp7.microsoft.com

Tracing route to corp7.microsoft.com [92.242.134.28]
over a maximum of 30 hops:

  1     *        *        *     Request timed out.
  2     *        *        *     Request timed out.
  3     *        *        *     Request timed out.
  4     *        *        *     Request timed out.
  5     *        *        *     Request timed out.
  6     *        *        *     Request timed out.
  7     *        *        *     Request timed out.
  8     *        *        *     Request timed out.
  9     *        *        *     Request timed out.
 10     *        *        *     Request timed out.
 11     *        *        *     Request timed out.
 12     *        *        *     Request timed out.
 13     *        *        *     Request timed out.
 14     *        *        *     Request timed out.
 15     *        *        *     Request timed out.
 16     *        *        *     Request timed out.
 17     *        *        *     Request timed out.
 18     *        *        *     Request timed out.
 19     *        *        *     Request timed out.
 20     *        *        *     Request timed out.
 21     *        *        *     Request timed out.
 22     *        *        *     Request timed out.
 23     *        *        *     Request timed out.
 24     *        *        *     Request timed out.
 25     *        *        *     Request timed out.
 26     *        *        *     Request timed out.
 27     *        *        *     Request timed out.
 28     *        *        *     Request timed out.
 29     *        *        *     Request timed out.
 30     *        *        *     Request timed out.

Trace complete.
```

f) Timeout in milliseconds.
```
C:\Windows\system32>TRACERT   -d   -4 -w 5000    www.google.es

Tracing route to www.google.es [216.58.201.131]
over a maximum of 30 hops:

  1    <1 ms    <1 ms    <1 ms  192.168.2.100
  2     7 ms     7 ms     7 ms  10.237.160.1
  3     8 ms    10 ms     6 ms  10.105.240.169
  4    13 ms    15 ms    18 ms  10.105.240.153
  5    32 ms    17 ms    12 ms  10.254.14.165
  6    18 ms    23 ms    16 ms  10.255.101.37
  7    18 ms    43 ms    29 ms  10.254.11.1
  8    49 ms    46 ms    47 ms  62.42.228.62
  9    20 ms    16 ms    39 ms  72.14.235.18
 10    14 ms    15 ms    15 ms  216.239.40.217
 11    17 ms    16 ms    18 ms  216.58.201.131

Trace complete.
```

> TRACERT -w number of milliseconds before each response, the default value 4000 (4 sec).

## STEP 6: Diagnose connection problems or latency. PathPing.

The pathping command is a route trace tool that combines features of the ping and tracert commands with additional information that neither of those tools provides. The pathping command sends packets to each router on the route to the final destination over a period of time and then computes results based on the packets returned from each hop. Since the command shows the level of packet loss on a specific link or router, it is easy to determine which routers or links might be causing network problems.

a) Forcing using IPv4 and try to reach a server by IP (8.8.8.8), it is a google server..
```
C:\Windows\system32>pathping  -4   8.8.8.8

Tracing route to google-public-dns-a.google.com [8.8.8.8]
over a maximum of 30 hops:
```

```
    0  SVR2017BSP.bspWeb.local [192.168.2.89]
    1  192.168.2.100
    2  10.237.160.1
    3  10.105.240.65
    4  10.254.14.165
    5  10.254.7.57
    6  10.254.10.150
    7  62.42.228.62.static.user.ono.com [62.42.228.62]
    8  72.14.234.231
    9  216.239.48.135
   10  google-public-dns-a.google.com [8.8.8.8]

Computing statistics for 250 seconds...
             Source to Here    This Node/Link
Hop  RTT    Lost/Sent = Pct   Lost/Sent = Pct  Address
 0                                              SVR2017BSP.bspWeb.local [192.168.2.89]
                              0/ 100 =  0%     |
 1   0ms    0/ 100 =  0%      0/ 100 =  0%     192.168.2.100
                              0/ 100 =  0%     |
 2   7ms    0/ 100 =  0%      0/ 100 =  0%     10.237.160.1
                              0/ 100 =  0%     |
 3   9ms    0/ 100 =  0%      0/ 100 =  0%     10.105.240.65
                              0/ 100 =  0%     |
 4   15ms   0/ 100 =  0%      0/ 100 =  0%     10.254.14.165
                              0/ 100 =  0%     |
 5   17ms   0/ 100 =  0%      0/ 100 =  0%     10.254.7.57
                              0/ 100 =  0%     |
 6   19ms   0/ 100 =  0%      0/ 100 =  0%     10.254.10.150
                              0/ 100 =  0%     |
 7   16ms   0/ 100 =  0%      0/ 100 =  0%     62.42.228.62.static.user.ono.com [62.42.228.62]
                              0/ 100 =  0%     |
 8   16ms   0/ 100 =  0%      0/ 100 =  0%     72.14.234.231
                              0/ 100 =  0%     |
 9   ---    100/ 100 =100%    100/ 100 =100%   216.239.48.135
                              0/ 100 =  0%     |
10   16ms   0/ 100 =  0%      0/ 100 =  0%     google-public-dns-a.google.com [8.8.8.8]

Trace complete.
```

b) Does not resolve addresses Host.

```
C:\Windows\system32>pathping -n  8.8.8.8

Tracing route to 8.8.8.8 over a maximum of 30 hops

    0  192.168.2.89
    1  192.168.2.100
    2  10.237.160.1
    3  10.105.240.65
    4  10.254.14.165
    5  10.254.7.57
    6  10.254.10.150
    7  62.42.228.62
    8  72.14.234.231
    9  216.239.48.135
   10  8.8.8.8

Computing statistics for 250 seconds...
             Source to Here    This Node/Link
Hop  RTT    Lost/Sent = Pct   Lost/Sent = Pct  Address
 0                                              192.168.2.89
                              0/ 100 =  0%     |
 1   0ms    0/ 100 =  0%      0/ 100 =  0%     192.168.2.100
                              0/ 100 =  0%     |
 2   6ms    0/ 100 =  0%      0/ 100 =  0%     10.237.160.1
                              0/ 100 =  0%     |
 3   8ms    0/ 100 =  0%      0/ 100 =  0%     10.105.240.65
                              0/ 100 =  0%     |
 4   15ms   0/ 100 =  0%      0/ 100 =  0%     10.254.14.165
                              0/ 100 =  0%     |
 5   17ms   0/ 100 =  0%      0/ 100 =  0%     10.254.7.57
                              0/ 100 =  0%     |
 6   18ms   0/ 100 =  0%      0/ 100 =  0%     10.254.10.150
                              0/ 100 =  0%     |
 7   15ms   0/ 100 =  0%      0/ 100 =  0%     62.42.228.62
                              0/ 100 =  0%     |
 8   16ms   0/ 100 =  0%      0/ 100 =  0%     72.14.234.231
                              0/ 100 =  0%     |
 9   ---    100/ 100 =100%    100/ 100 =100%   216.239.48.135
                              0/ 100 =  0%     |
10   16ms   0/ 100 =  0%      0/ 100 =  0%     8.8.8.8

Trace complete.
```

> PATHPING -h number of hops between 1-255, default 30

c) Set the maximum number of hops.

```
C:\Windows\system32>pathping -h 100  8.8.8.8

Tracing route to google-public-dns-a.google.com [8.8.8.8]
over a maximum of 100 hops:
    0  i7-PC [192.168.1.99]
    1  192.168.1.1
    2  192.168.144.1
    3   *        *        *
Computing statistics for 50 seconds...
             Source to Here    This Node/Link
Hop  RTT    Lost/Sent = Pct   Lost/Sent = Pct  Address
 0                                              i7-PC [192.168.1.99]
```

> PATHPING -p second delay between each ping between 1-255, default 250 (1/4 second)

```
    1   0ms      0/ 100 =   0%     0/ 100 =   0%  192.168.1.1
                                 100/ 100 =100%   |
    2   ---    100/ 100 =100%     0/ 100 =   0%  192.168.144.1

Trace complete.
C:\Windows\system32>pathping -h 100  8.8.8.8

Tracing route to google-public-dns-a.google.com [8.8.8.8]
over a maximum of 100 hops:
  0  SVR2017BSP.bspWeb.local [192.168.2.89]
  1  192.168.2.100
  2  10.237.160.1
  3  10.105.240.65
  4  10.254.14.165
  5  10.254.7.57
  6  10.254.10.150
  7  62.42.228.62.static.user.ono.com [62.42.228.62]
  8  72.14.234.231
  9  216.239.48.135
 10  google-public-dns-a.google.com [8.8.8.8]

Computing statistics for 250 seconds...
            Source to Here    This Node/Link
Hop  RTT   Lost/Sent = Pct   Lost/Sent = Pct  Address
  0                                            SVR2017BSP.bspWeb.local [192.168.2.89]
                                0/ 100 =   0%  |
  1   0ms     0/ 100 =   0%     0/ 100 =   0%  192.168.2.100
                                0/ 100 =   0%  |
  2   6ms     0/ 100 =   0%     0/ 100 =   0%  10.237.160.1
                                0/ 100 =   0%  |
  3   8ms     0/ 100 =   0%     0/ 100 =   0%  10.105.240.65
                                0/ 100 =   0%  |
  4  17ms     0/ 100 =   0%     0/ 100 =   0%  10.254.14.165
                                0/ 100 =   0%  |
  5  17ms     0/ 100 =   0%     0/ 100 =   0%  10.254.7.57
                                0/ 100 =   0%  |
  6  20ms     0/ 100 =   0%     0/ 100 =   0%  10.254.10.150
                                0/ 100 =   0%  |
  7  15ms     0/ 100 =   0%     0/ 100 =   0%  62.42.228.62.static.user.ono.com [62.42.228.62]
                                0/ 100 =   0%  |
  8  16ms     0/ 100 =   0%     0/ 100 =   0%  72.14.234.231
                                0/ 100 =   0%  |
  9  ---    100/ 100 =100%    100/ 100 =100%   216.239.48.135
                                0/ 100 =   0%  |
 10  16ms     0/ 100 =   0%     0/ 100 =   0%  google-public-dns-a.google.com [8.8.8.8]

Trace complete.
```

d) Number of seconds to wait between each ping.
```
C:\Windows\system32>pathping  -p 5  8.8.8.8

Tracing route to google-public-dns-a.google.com [8.8.8.8]
over a maximum of 100 hops:
  0  i7-PC [192.168.1.99]
  1  192.168.1.1
  2  192.168.144.1
  3   *        *        *
Computing statistics for 250 seconds...
            Source to Here    This Node/Link
Hop  RTT   Lost/Sent = Pct   Lost/Sent = Pct  Address
  0                                            i7-PC [192.168.1.99]
                                0/ 100 =   0%  |
  1   0ms     0/ 100 =   0%     0/ 100 =   0%  192.168.1.1
                                100/ 100 =100% |
  2   ---   100/ 100 =100%      0/ 100 =   0%  192.168.144.1

Trace complete.
```

> PATHPING -q number of queries by leaps between 1-255, default 100.

e) Establish a specific IP address as source.
```
C:\Windows\system32>pathping -i 192.168.1.120  8.8.8.8

Tracing route to google-public-dns-a.google.com [8.8.8.8]
over a maximum of 30 hops:

The option ? is only supported for IPv6.
```

f) Use only IPv6, IP and then first name. Name resolves.
```
C:\Windows\system32> pathping -6 8.8.8.8
Unable to resolve target system name 8.8.8.8.

C:\Windows\system32> pathping -6 www.google.es

Tracing route to www.google.es [2a00:1450:4003:804::2003]
over a maximum of 30 hops:
  0  i7-PC [fe80::b9ca:2a45:541d:2e0f%9]
  1   *        *        *
Computing statistics for 0 seconds...
            Source to Here    This Node/Link
Hop  RTT   Lost/Sent = Pct   Lost/Sent = Pct  Address
  0                                            i7-PC [fe80::b9ca:2a45:541d:2e0f%9]

Trace complete.
```

g) Establish an IPv4 source address and uses only IPv6.
```
C:\Windows\system32>pathping -i 192.168.1.120 -6 8.8.8.8
Unable to resolve target system name 8.8.8.8.
```

h) Establish a number of queries per jump.
```
C:\Windows\system32> pathping -q 255 -6 www.google.es

Tracing route to www.google.es [2a00:1450:4003:804::2003]
over a maximum of 30 hops:
  0  i7-PC [fe80::b9ca:2a45:541d:2e0f%9]
  1  *       *        *
Computing statistics for 0 seconds...
            Source to Here   This Node/Link
Hop  RTT    Lost/Sent = Pct  Lost/Sent = Pct  Address
  0                                           i7-PC [fe80::b9ca:2a45:541d:2e0f%9]

Trace complete.
```

g) The number of queries per hop seconds time waiting for each ping and milliseconds is-pear before each response is established. This is feasible with IPv4 but not IPv6.
```
C:\Windows\system32> pathping -q 255  -p 100  -w 50    www.google.es

Tracing route to www.google.es [216.58.201.131]
over a maximum of 30 hops:
  0  i7-PC [192.168.1.99]
  1  192.168.1.1
  2  192.168.144.1
  3  *       *        *
Computing statistics for 51 seconds...
            Source to Here   This Node/Link
Hop  RTT    Lost/Sent = Pct  Lost/Sent = Pct  Address
  0                                           i7-PC [192.168.1.99]
                              0/ 255 =  0%    |
  1   0ms    0/ 255 =  0%     0/ 255 =  0%    192.168.1.1
                            255/ 255 =100%    |
  2   ---  255/ 255 =100%     0/ 255 =  0%    192.168.144.1

Trace complete.
```

> PATHPING -w number of milliseconds before each response, the default value 3000 (3sec).

h) Timeouts response to very high jumps. In principle I have found no limit on the timeout parameter (tested up to 256 trillion).
```
C:\Windows\system32>PATHPING -q 255  -p 100  -w 256000  www.google.es

Tracing route to www.google.es [216.58.201.131]
over a maximum of 30 hops:
  0  SVR2017BSP.bspWeb.local [192.168.2.89]
  1  192.168.2.100
  2  10.237.160.1
  3  10.105.240.169
  4  10.105.240.153
  5  10.254.14.165
  6  10.255.101.37
  7  10.254.11.1
  8  62.42.228.62.static.user.ono.com [62.42.228.62]
  9  72.14.235.18
 10  216.239.40.217
 11  mad06s25-in-f131.1e100.net [216.58.201.131]

Computing statistics for 280 seconds...
            Source to Here   This Node/Link
Hop  RTT    Lost/Sent = Pct  Lost/Sent = Pct  Address
  0                                           SVR2017BSP.bspWeb.local [192.168.2.89]
                              0/ 255 =  0%    |
  1   0ms    0/ 255 =  0%     0/ 255 =  0%    192.168.2.100
                              0/ 255 =  0%    |
  2   7ms    0/ 255 =  0%     0/ 255 =  0%    10.237.160.1
                              0/ 255 =  0%    |
  3   8ms    0/ 255 =  0%     0/ 255 =  0%    10.105.240.169
                              0/ 255 =  0%    |
  4   8ms    0/ 255 =  0%     0/ 255 =  0%    10.105.240.153
                              0/ 255 =  0%    |
  5  16ms    0/ 255 =  0%     0/ 255 =  0%    10.254.14.165
                              0/ 255 =  0%    |
  6  19ms    0/ 255 =  0%     0/ 255 =  0%    10.255.101.37
                              0/ 255 =  0%    |
  7  20ms    0/ 255 =  0%     0/ 255 =  0%    10.254.11.1
                              0/ 255 =  0%    |
  8  15ms    0/ 255 =  0%     0/ 255 =  0%    62.42.228.62.static.user.ono.com [62.42.228.62]
                              0/ 255 =  0%    |
  9  18ms    0/ 255 =  0%     0/ 255 =  0%    72.14.235.18
                              0/ 255 =  0%    |
 10  15ms    0/ 255 =  0%     0/ 255 =  0%    216.239.40.217
                              0/ 255 =  0%    |
 11  16ms    0/ 255 =  0%     0/ 255 =  0%    mad06s25-in-f131.1e100.net [216.58.201.131]

Trace complete.
```

# PRACTICE 10: Test the DNS server.

DESCRIPTION:

DNS stands for Domain Name System (domain names) and is a technology based on a database that is used to resolve names on the network, ie, to know the IP address of the machine where it is housed-do the domain you want to access.

When a computer is connected to a network (either Internet or a home network) is assigned an IP address, associated to a MAC-da. If there is a very large volume of IPs becomes impossible, so there are domains and DNS-hazards arising, for translat.

The DNS is a system that serves to translate the names on the network, and is composed of three parts with distinct functions.

Client DNS is installed on the client (IP written in DNS) and performs name resolution requests to the DNS-res served.

DNS server: they are answering requests and resolve the names through a tree structured system. DNS addresses that we put in the connection settings are the addresses of the DNS servers.

Areas of authority are or groups of servers that are assigned to solve a particular set of domains (such as .com or .org).

## How does it work?

Name resolution uses a tree structure, whereby the various DNS servers authority areas are responsible for resolving the addresses in your area, and it is requested by another server who believe they know the address.

## STEP 1: Access the NSLOOKUP utility

It can be run directly from the command line or from within the application, which is standard.

a) Check operation of DNS SERVER.

```
NSLOOKUP
C:\Windows\system32>ping 192.168.2.100

Pinging 192.168.2.100 with 32 bytes of data:
Reply from 192.168.2.100: bytes=32 time=1ms TTL=64
Reply from 192.168.2.100: bytes=32 time=3ms TTL=64
Reply from 192.168.2.100: bytes=32 time=1ms TTL=64
Reply from 192.168.2.100: bytes=32 time<1ms TTL=64

Ping statistics for 192.168.2.100:
    Packets: Sent = 4, Received = 4, Lost = 0 (0% loss),
Approximate round trip times in milli-seconds:
    Minimum = 0ms, Maximum = 3ms, Average = 1ms

C:\Windows\system32>ping 192.168.2.89

Pinging 192.168.2.89 with 32 bytes of data:
Reply from 192.168.2.89: bytes=32 time<1ms TTL=128
Reply from 192.168.2.89: bytes=32 time<1ms TTL=128
Reply from 192.168.2.89: bytes=32 time<1ms TTL=128
Reply from 192.168.2.89: bytes=32 time<1ms TTL=128

Ping statistics for 192.168.2.89:
    Packets: Sent = 4, Received = 4, Lost = 0 (0% loss),
Approximate round trip times in milli-seconds:
    Minimum = 0ms, Maximum = 0ms, Average = 0ms

C:\Windows\system32>nslookup
Default Server:  localhost
Address:  127.0.0.1

> svr2017BSP
Server:  localhost
Address:  127.0.0.1

Name:    svr2017BSP.bspWeb.local
Addresses:  192.168.5.240
            192.168.2.89
            192.168.1.89

> bspWeb.local
Server:  localhost
Address:  127.0.0.1

Name:    bspWeb.local
Addresses:  192.168.5.240
            192.168.1.89
            192.168.2.89
```

b) Help.
- b.1) Completed within the utility nslookup. It is typed
  help
- b.2) Aid allocation.
  set all

c) Select a way to query the DNS.
> set type=NS

```
> google.es
Server: localhost
Address: 127.0.0.1

Non-authoritative answer:
google.es      nameserver = ns4.google.com
google.es      nameserver = ns3.google.com
google.es      nameserver = ns2.google.com
google.es      nameserver = ns1.google.com

ns4.google.com  internet address = 216.239.38.10
ns3.google.com  internet address = 216.239.36.10
ns2.google.com  internet address = 216.239.34.10
ns1.google.com  internet address = 216.239.32.10
```

d) Refresh the DNS. Before entering loopback.
   IPCONFIG  /FLUSHDNS

```
Windows IP Configuration

Successfully flushed the DNS Resolver Cache.
```
View all the information on the card with IPCONFIG.
   IPCONFIG  /ALL
```
C:\Windows\system32>IPCONFIG      /ALL

Windows IP Configuration

    Host Name . . . . . . . . . . . . : SVR2017BSP
    Primary Dns Suffix  . . . . . . . : bspWeb.local
    Node Type . . . . . . . . . . . . : Hybrid
    IP Routing Enabled. . . . . . . . : No
    WINS Proxy Enabled. . . . . . . . : No
    DNS Suffix Search List. . . . . . : bspWeb.local

Ethernet adapter Ethernet:

    Connection-specific DNS Suffix  . :
    Description . . . . . . . . . . . : Microsoft Hyper-V Network Adapter
    Physical Address. . . . . . . . . : 00-15-5D-01-63-06
    DHCP Enabled. . . . . . . . . . . : No
    Autoconfiguration Enabled . . . . : Yes
    IPv4 Address. . . . . . . . . . . : 192.168.1.89(Preferred)
    Subnet Mask . . . . . . . . . . . : 255.255.255.0
    IPv4 Address. . . . . . . . . . . : 192.168.2.89(Preferred)
    Subnet Mask . . . . . . . . . . . : 255.255.255.0
    IPv4 Address. . . . . . . . . . . : 192.168.5.240(Preferred)
    Subnet Mask . . . . . . . . . . . : 255.255.255.0
    Default Gateway . . . . . . . . . : 192.168.2.100
                                        192.168.1.1
    DNS Servers . . . . . . . . . . . : 127.0.0.1
    NetBIOS over Tcpip. . . . . . . . : Enabled

Tunnel adapter isatap.{0D7F8ED4-1EC1-4BB3-A058-D0C317F404A5}:

    Media State . . . . . . . . . . . : Media disconnected
    Connection-specific DNS Suffix  . :
    Description . . . . . . . . . . . : Microsoft ISATAP Adapter #2
    Physical Address. . . . . . . . . : 00-00-00-00-00-00-00-E0
    DHCP Enabled. . . . . . . . . . . : No
    Autoconfiguration Enabled . . . . : Yes
```
e) Consultation by associating a name with a name server.
```
C:\Windows\system32>nslookup  8.8.8.8
Server:  localhost
Address:  127.0.0.1

Name:    google-public-dns-a.google.com
Address:  8.8.8.8
```
f) Ask for Alias.
```
C:\Windows\system32>nslookup
> set type=CNAME
> google.es
Server:  localhost
Address:  127.0.0.1

google.es
        primary name server = ns4.google.com
```

```
            responsible mail addr = dns-admin.google.com
            serial  = 133162154
            refresh = 900 (15 mins)
            retry   = 900 (15 mins)
            expire  = 1800 (30 mins)
            default TTL = 60 (1 min)
> 8.8.8.8
Server:  localhost
Address:  127.0.0.1

*** localhost can't find 8.8.8.8: Non-existent domain
```

g) Check for IPv4 address.

```
> set type=A
> 8.8.8.8
Server:  localhost
Address:  127.0.0.1

Name:     google-public-dns-a.google.com
Address:  8.8.8.8

> google.es
Server:  localhost
Address:  127.0.0.1

Non-authoritative answer:
Name:     google.es
Address:  216.58.201.131

> google.com
Server:  localhost
Address:  127.0.0.1

Non-authoritative answer:
Name:     google.com
Address:  216.58.211.206
```

**Web servers that allow poll: PING, TRACERT, TRACEROUTER, NSLOOKUP, TELNET, FTP, TFTP, PUERTOS,...**
http://www.kloth.net/
http://dig-nslookup.nmonitoring.com
http://toolbox.googleapps.net/apps/dig
http://network-tools.com/nslook/
http://subnetonline.com/pages/subnet-cola
http://ping.eu/nslookup
http://mxtoolbox.com/SuperTool.aspx

h) Query by IPv6 address.

```
> set type=AAAA
> google.es
Server:  localhost
Address:  127.0.0.1

Non-authoritative answer:
Name:     google.es
Address:  2a00:1450:4003:804::2003
```

i) Consultation for IPv4 and IPv6 address.

```
> set type=A+AAAA
> google.es
Server:  localhost
Address:  127.0.0.1

Non-authoritative answer:
Name:      google.es
Addresses: 2a00:1450:4003:804::2003
           216.58.201.131

> 8.8.8.8
Server:  localhost
Address:  127.0.0.1

Name:     google-public-dns-a.google.com
Address:  8.8.8.8
```

j) Ask for a mail server.

Mail servers use this information to find where to redirect emails sent to a particular address.

```
> set type=MX
> google.es
Server:  localhost
Address:  127.0.0.1

Non-authoritative answer:
google.es      MX preference = 40, mail exchanger = alt3.aspmx.l.google.com
google.es      MX preference = 50, mail exchanger = alt4.aspmx.l.google.com
google.es      MX preference = 30, mail exchanger = alt2.aspmx.l.google.com
google.es      MX preference = 20, mail exchanger = alt1.aspmx.l.google.com
google.es      MX preference = 10, mail exchanger = aspmx.l.google.com

alt3.aspmx.l.google.com internet address = 64.233.189.26
alt3.aspmx.l.google.com AAAA IPv6 address = 2404:6800:4008:c07::1b
alt4.aspmx.l.google.com internet address = 173.194.72.27
alt4.aspmx.l.google.com AAAA IPv6 address = 2404:6800:4008:c01::1a
alt2.aspmx.l.google.com internet address = 74.125.68.27
alt2.aspmx.l.google.com AAAA IPv6 address = 2404:6800:4003:c02::1b
```

```
        alt1.aspmx.l.google.com  internet address = 173.194.221.27
        alt1.aspmx.l.google.com  AAAA IPv6 address = 2a00:1450:4010:c0a::1b
        aspmx.l.google.com       internet address = 64.233.184.27
        aspmx.l.google.com       AAAA IPv6 address = 2a00:1450:400c:c07::1a
```
k) Ask for a mail server.
```
        > set type=hinfo
        > google.es
        Server:  localhost
        Address: 127.0.0.1

        google.es
                primary name server = ns2.google.com
                responsible mail addr = dns-admin.google.com
                serial   = 133162154
                refresh  = 900 (15 mins)
                retry    = 900 (15 mins)
                expire   = 1800 (30 mins)
                default TTL = 60 (1 min)
```
l) Ask for a mail server.
```
        > set type=soa
        > google.es
        Server:  localhost
        Address: 127.0.0.1

        Non-authoritative answer:
        google.es
                primary name server = ns1.google.com
                responsible mail addr = dns-admin.google.com
                serial   = 133162154
                refresh  = 900 (15 mins)
                retry    = 900 (15 mins)
                expire   = 1800 (30 mins)
                default TTL = 60 (1 min)

        ns1.google.com  internet address = 216.239.32.10
```
m) Ask for a mail server.
```
        > set type=PTR
        > google.es
        Server:  localhost
        Address: 127.0.0.1

        google.es
                primary name server = ns3.google.com
                responsible mail addr = dns-admin.google.com
                serial   = 133162154
                refresh  = 900 (15 mins)
                retry    = 900 (15 mins)
                expire   = 1800 (30 mins)
                default TTL = 60 (1 min)
        > 8.8.8.8
        Server:  localhost
        Address: 127.0.0.1

        Non-authoritative answer:
        8.8.8.8.in-addr.arpa    name = google-public-dns-a.google.com
```
n) Check for mail server.
```
        > set type=ANY
        > google.es
        Server:  localhost
        Address: 127.0.0.1

        Non-authoritative answer:
        google.es        internet address = 216.58.201.131
        google.es        nameserver = ns2.google.com
        google.es        nameserver = ns1.google.com
        google.es        nameserver = ns4.google.com
        google.es        nameserver = ns3.google.com
        google.es
                primary name server = ns1.google.com
                responsible mail addr = dns-admin.google.com
                serial   = 133162154
                refresh  = 900 (15 mins)
                retry    = 900 (15 mins)
                expire   = 1800 (30 mins)
                default TTL = 60 (1 min)
        google.es        MX preference = 50, mail exchanger = alt4.aspmx.l.google.com
        google.es        MX preference = 30, mail exchanger = alt2.aspmx.l.google.com
        google.es        MX preference = 20, mail exchanger = alt1.aspmx.l.google.com
        google.es        MX preference = 10, mail exchanger = aspmx.l.google.com
        google.es        MX preference = 40, mail exchanger = alt3.aspmx.l.google.com
        google.es        text =
```

```
            "v=spf1 -all"
google.es          AAAA IPv6 address = 2a00:1450:4003:804::2003

ns2.google.com   internet address = 216.239.34.10
ns1.google.com   internet address = 216.239.32.10
ns4.google.com   internet address = 216.239.38.10
ns3.google.com   internet address = 216.239.36.10
alt4.aspmx.l.google.com internet address = 173.194.72.27
alt4.aspmx.l.google.com AAAA IPv6 address = 2404:6800:4008:c01::1a
alt2.aspmx.l.google.com internet address = 74.125.68.27
alt2.aspmx.l.google.com AAAA IPv6 address = 2404:6800:4003:c02::1b
alt1.aspmx.l.google.com internet address = 173.194.221.27
            "v=spf1 -all"
google.es          MX preference = 50, mail exchanger = alt4.aspmx.l.google.com
google.es          nameserver = ns4.google.com
```

# PRACTICE 11: Routing Tables. ROUTER

DESCRIPCIÓN:

### The routing table

The routing table is a table of connections between the management team and the destination node through which the router should send the message, it is necessary to store the complete IP address of the computer.

The routing table contains pairs of IP addresses.

In any communication there is a sender and a receiver, we start from the sender and receiver belong to the same network, we talk about direct delivery. But if there is at least one router between the sending and the receiving, we talk about indirect delivery.

In the case of an indirect delivery, the role of the router and, in particular, the routing table is very matter-you. The operation of a router is determined by the way in which this routing table (static or dynamic) is created.

- The administrator manually enter the routing table, this is called routing estáti-co (suitable for small networks).
- The router builds its own routing tables, using the information received through the protoco-routing, this is called dynamic routing.

There are different levels of routers which operate with different protocols:

- Routers node: are the main routers that establish the connection between different networks.
- External Routers: have the ability to establish the connection to autonomous networks together. Its operation is based on the protocol called EGP (Exterior Gateway Protocol).
- Internal Routers sets the routing information within an autonomous network. They exchange information using protocols called IGP (Interior Gateway Protocol), such as RIP and OSPF.

## STEP 1: View routing tables.

a) Display the routing tables.

```
C:\Windows\system32>route print
===========================================================================
Interface List
 12...00 15 5d 01 63 06 ......Microsoft Hyper-V Network Adapter
  1...........................Software Loopback Interface 1
 15...00 00 00 00 00 00 00 e0 Microsoft ISATAP Adapter #2
===========================================================================

IPv4 Route Table
===========================================================================
Active Routes:
Network Destination        Netmask          Gateway       Interface  Metric
          0.0.0.0          0.0.0.0      192.168.2.100     192.168.1.89    276
          0.0.0.0          0.0.0.0        192.168.1.1     192.168.1.89    276
        127.0.0.0        255.0.0.0         On-link         127.0.0.1      306
        127.0.0.1  255.255.255.255         On-link         127.0.0.1      306
  127.255.255.255  255.255.255.255         On-link         127.0.0.1      306
      192.168.1.0    255.255.255.0         On-link       192.168.1.89     276
     192.168.1.89  255.255.255.255         On-link       192.168.1.89     276
    192.168.1.255  255.255.255.255         On-link       192.168.1.89     276
      192.168.2.0    255.255.255.0         On-link       192.168.1.89     276
     192.168.2.89  255.255.255.255         On-link       192.168.1.89     276
    192.168.2.255  255.255.255.255         On-link       192.168.1.89     276
      192.168.5.0    255.255.255.0         On-link       192.168.1.89     276
    192.168.5.240  255.255.255.255         On-link       192.168.1.89     276
    192.168.5.255  255.255.255.255         On-link       192.168.1.89     276
        224.0.0.0        240.0.0.0         On-link         127.0.0.1      306
        224.0.0.0        240.0.0.0         On-link       192.168.1.89     276
  255.255.255.255  255.255.255.255         On-link         127.0.0.1      306
  255.255.255.255  255.255.255.255         On-link       192.168.1.89     276
===========================================================================
Persistent Routes:
  Network Address          Netmask  Gateway Address  Metric
          0.0.0.0          0.0.0.0    192.168.2.100  Default
          0.0.0.0          0.0.0.0      192.168.1.1  Default
===========================================================================

IPv6 Route Table
===========================================================================
Active Routes:
 If Metric Network Destination      Gateway
  1    306 ::1/128                  On-link
  1    306 ff00::/8                 On-link
===========================================================================
Persistent Routes:
  None
```

b) Display IPv4 routing tables.

```
C:\Windows\system32>route print -4
===========================================================================
```

```
Interface List
 12...00 15 5d 01 63 06 ......Microsoft Hyper-V Network Adapter
  1...........................Software Loopback Interface 1
 15...00 00 00 00 00 00 00 e0 Microsoft ISATAP Adapter #2
===========================================================================

IPv4 Route Table
===========================================================================
Active Routes:
Network Destination        Netmask          Gateway       Interface  Metric
          0.0.0.0          0.0.0.0      192.168.2.100    192.168.1.89    276
          0.0.0.0          0.0.0.0        192.168.1.1    192.168.1.89    276
        127.0.0.0        255.0.0.0          On-link         127.0.0.1    306
        127.0.0.1  255.255.255.255          On-link         127.0.0.1    306
  127.255.255.255  255.255.255.255          On-link         127.0.0.1    306
      192.168.1.0    255.255.255.0          On-link      192.168.1.89    276
     192.168.1.89  255.255.255.255          On-link      192.168.1.89    276
    192.168.1.255  255.255.255.255          On-link      192.168.1.89    276
      192.168.2.0    255.255.255.0          On-link      192.168.1.89    276
     192.168.2.89  255.255.255.255          On-link      192.168.1.89    276
    192.168.2.255  255.255.255.255          On-link      192.168.1.89    276
      192.168.5.0    255.255.255.0          On-link      192.168.1.89    276
    192.168.5.240  255.255.255.255          On-link      192.168.1.89    276
    192.168.5.255  255.255.255.255          On-link      192.168.1.89    276
        224.0.0.0        240.0.0.0          On-link         127.0.0.1    306
        224.0.0.0        240.0.0.0          On-link      192.168.1.89    276
  255.255.255.255  255.255.255.255          On-link         127.0.0.1    306
  255.255.255.255  255.255.255.255          On-link      192.168.1.89    276
===========================================================================
Persistent Routes:
  Network Address          Netmask  Gateway Address  Metric
          0.0.0.0          0.0.0.0    192.168.2.100  Default
          0.0.0.0          0.0.0.0      192.168.1.1  Default
===========================================================================
```

c) Clears the routing tables of all gateway entries.
   ROUTE -F
```
C:\Windows\system32> route  -f
 OK!

C:\Windows\system32> route print
C:\Windows\system32>route print
===========================================================================
Interface List
 12...00 15 5d 01 63 06 ......Microsoft Hyper-V Network Adapter
  1...........................Software Loopback Interface 1
 15...00 00 00 00 00 00 00 e0 Microsoft ISATAP Adapter #2
===========================================================================

IPv4 Route Table
===========================================================================
Active Routes:
  None
Persistent Routes:
  None

IPv6 Route Table
===========================================================================
Active Routes:
  None
Persistent Routes:
  None
```

d) Add a route to the route table.
   ROUTE  ADD 192.168.2.50 MASK  255.255.255.0  192.168.2.100
   ROUTE  ADD 192.168.2.50 MASK  255.255.255.0  192.168.2.100  METRIC 3  IF 0

## STEP 2: Display and modify data translation table of IP addresses to MAC addresses (ARP table)

a) Visualize the ARP table for each of the interfaces.
```
C:\Windows\system32>arp -a

Interface: 192.168.1.89 --- 0xc
  Internet Address      Physical Address      Type
  192.168.2.100         dc-53-7c-60-58-5a     dynamic
  224.0.0.2             01-00-5e-00-00-02     static
  224.0.0.22            01-00-5e-00-00-16     static
  224.0.0.113           01-00-5e-00-00-71     static
  224.0.0.251           01-00-5e-00-00-fb     static
  224.0.0.252           01-00-5e-00-00-fc     static
  239.255.255.250       01-00-5e-7f-ff-fa     static
```

b) Same as option a).
```
C:\Windows\system32>arp -G

Interface: 192.168.1.89 --- 0xc
  Internet Address      Physical Address      Type
  192.168.2.100         dc-53-7c-60-58-5a     dynamic
  224.0.0.2             01-00-5e-00-00-02     static
  224.0.0.22            01-00-5e-00-00-16     static
  224.0.0.113           01-00-5e-00-00-71     static
  224.0.0.251           01-00-5e-00-00-fb     static
  224.0.0.252           01-00-5e-00-00-fc     static
  239.255.255.250       01-00-5e-7f-ff-fa     static
```
c) Add a specific entry to the ARP table.
```
C:\Windows\system32>ARP -S 145.5.2.2 00-00-01-aa-ff-aa

C:\Windows\system32>arp -A

Interface: 192.168.1.89 --- 0xc
  Internet Address      Physical Address      Type
  145.5.2.2             00-00-01-aa-ff-aa     static
  192.168.2.100         dc-53-7c-60-58-5a     dynamic
  224.0.0.2             01-00-5e-00-00-02     static
  224.0.0.22            01-00-5e-00-00-16     static
  224.0.0.113           01-00-5e-00-00-71     static
  224.0.0.251           01-00-5e-00-00-fb     static
  224.0.0.252           01-00-5e-00-00-fc     static
  239.255.255.250       01-00-5e-7f-ff-fa     static
```
d) Removes a specific entry from the ARP table.
```
C:\Windows\system32>arp -D
The ARP entry deletion failed: The parameter is incorrect.


C:\Windows\system32>arp -A

Interface: 192.168.1.89 --- 0xc
  Internet Address      Physical Address      Type
  145.5.2.2             00-00-01-aa-ff-aa     static
  192.168.2.100         dc-53-7c-60-58-5a     dynamic
  224.0.0.2             01-00-5e-00-00-02     static
  224.0.0.22            01-00-5e-00-00-16     static
  224.0.0.113           01-00-5e-00-00-71     static
  224.0.0.251           01-00-5e-00-00-fb     static
  224.0.0.252           01-00-5e-00-00-fc     static
  239.255.255.250       01-00-5e-7f-ff-fa     static
```
e) Display of detailed information detailed way. all invalid entries and entries will be displayed on the loopback interface.
```
C:\Windows\system32>arp -G -V

Interface: 127.0.0.1 --- 0x1
  Internet Address      Physical Address      Type
  127.0.0.1                                   static
  224.0.0.2                                   static
  224.0.0.22                                  static
  224.0.0.113                                 static
  224.0.0.251                                 static
  239.255.255.250                             static

Interface: 192.168.1.89 --- 0xc
  Internet Address      Physical Address      Type
  127.0.0.1             00-00-00-00-00-00     invalid
  145.5.2.2             00-00-01-aa-ff-aa     static
  192.168.1.1           00-00-00-00-00-00     invalid
  192.168.2.100         dc-53-7c-60-58-5a     dynamic
  224.0.0.2             01-00-5e-00-00-02     static
  224.0.0.22            01-00-5e-00-00-16     static
  224.0.0.113           01-00-5e-00-00-71     static
  224.0.0.251           01-00-5e-00-00-fb     static
  224.0.0.252           01-00-5e-00-00-fc     static
  239.255.255.250       01-00-5e-7f-ff-fa     static
```
f) Displays the ARP entries for the network interface specified by if_addr.
```
ARP    -N      8.8.8.8
```

# ATTACHMENTS: RED commands and AD DS

## ARP
Displays and modifies the IP-to-Physical address translation tables used by address resolution protocol (ARP).

```
ARP -s inet_addr eth_addr [if_addr]
ARP -d inet_addr [if_addr]
ARP -a [inet_addr] [-N if_addr] [-v]
```

| | |
|---|---|
| `-a` | Displays current ARP entries by interrogating the current protocol data. If inet_addr is specified, the IP and Physical addresses for only the specified computer are displayed. If more than one network interface uses ARP, entries for each ARP table are displayed. |
| `-g` | Same as -a. |
| `-v` | Displays current ARP entries in verbose mode. All invalid entries and entries on the loop-back interface will be shown. |
| `inet_addr` | Specifies an internet address. |
| `-N if_addr` | Displays the ARP entries for the network interface specified by if_addr. |
| `-d` | Deletes the host specified by inet_addr. inet_addr may be wildcarded with * to delete all hosts. |
| `-s` | Adds the host and associates the Internet address inet_addr with the Physical address eth_addr. The Physical address is given as 6 hexadecimal bytes separated by hyphens. The entry is permanent. |
| `eth_addr` | Specifies a physical address. |
| `if_addr` | If present, this specifies the Internet address of the interface whose address translation table should be modified. If not present, the first applicable interface will be used. |

**Example:**
```
> arp -s 157.55.85.212  00-aa-00-62-c6-09       .... Adds a static entry.
> arp -a                                         .... Displays the arp table.
```

## ICACLS
Stores the DACLs for the files and folders that match the name into aclfile for later use with /restore. Note that SACLs, owner, or integrity labels are not saved.

```
ICACLS name /save aclfile [/T] [/C] [/L] [/Q]
```

ICACLS directory [/substitute SidOld SidNew [...]] /restore aclfile [/C] [/L] [/Q] applies the stored DACLs to files in directory.

ICACLS name /setowner user [/T] [/C] [/L] [/Q]  changes the owner of all matching names. This option does not force a change of ownership; use the takeown.exe utility for that purpose.

ICACLS name /findsid Sid [/T] [/C] [/L] [/Q] finds all matching names that contain an ACL explicitly mentioning Sid.

ICACLS name /verify [/T] [/C] [/L] [/Q] finds all files whose ACL is not in canonical form or whose lengths are inconsistent with ACE counts.

ICACLS name /reset [/T] [/C] [/L] [/Q]   replaces ACLs with default inherited ACLs for all matching files.

```
ICACLS name [/grant[:r] Sid:perm[...]]
      [/deny Sid:perm [...]]
      [/remove[:g|:d]] Sid[...]] [/T] [/C] [/L] [/Q]
      [/setintegritylevel Level:policy[...]]
```

**/grant[:r] Sid:perm** grants the specified user access rights. With :r, the permissions replace any previously granted explicit permissions.
Without :r, the permissions are added to any previously granted explicit permissions.

**/deny Sid:perm** explicitly denies the specified user access rights.
An explicit deny ACE is added for the stated permissions and the same permissions in any explicit grant are removed.

**/remove[:[g|d]]** Sid removes all occurrences of Sid in the ACL. With
:g, it removes all occurrences of granted rights to that Sid. With
:d, it removes all occurrences of denied rights to that Sid.

**/setintegritylevel [(CI)(OI)]**Level explicitly adds an integrity ACE to all matching files. The level is to be specified as one of:
- **L[ow]**
- **M[edium]**
- **H[igh]**

Inheritance options for the integrity ACE may precede the level and are applied only to directories.

**/inheritance:e|d|r**
- **e** - enables inheritance
- **d** - disables inheritance and copy the ACEs
- **r** - remove all inherited ACEs

**Note:**
Sids may be in either numerical or friendly name form. If a numerical form is given, affix a * to the start of the SID.

**/T**   indicates that this operation is performed on all matching files/directories below the directories specified in the name.

```
/C      indicates that this operation will continue on all file errors. Error messages will still be dis-
        played.
/L      indicates that this operation is performed on a symbolic link itself versus its target.
/Q      indicates that icacls should suppress success messages.
ICACLS preserves the canonical ordering of ACE entries:
        Explicit denials
        Explicit grants
        Inherited denials
        Inherited grants
perm is a permission mask and can be specified in one of two forms:
    a sequence of simple rights:
            N - no access
            F - full access
            M - modify access
            RX - read and execute access
            R - read-only access
            W - write-only access
            D - delete access
    a comma-separated list in parentheses of specific rights:
            DE - delete
            RC - read control
            WDAC - write DAC
            WO - write owner
            S - synchronize
            AS - access system security
            MA - maximum allowed
            GR - generic read
            GW - generic write
            GE - generic execute
            GA - generic all
            RD - read data/list directory
            WD - write data/add file
            AD - append data/add subdirectory
            REA - read extended attributes
            WEA - write extended attributes
            X - execute/traverse
            DC - delete child
            RA - read attributes
            WA - write attributes
    inheritance rights may precede either form and are applied only to directories:
            (OI) - object inherit
            (CI) - container inherit
            (IO) - inherit only
            (NP) - don't propagate inherit
            (I) - permission inherited from parent container

Examples:
                    icacls c:\windows\* /save AclFile /T
Will save the ACLs for all files under c:\windows and its subdirectories to AclFile.
                    icacls c:\windows\ /restore AclFile
Will restore the Acls for every file within AclFile that exists in c:\windows and its subdirectories.
                    icacls file /grant Administrator:(D,WDAC)
Will grant the user Administrator Delete and Write DAC permissions to file.
                    icacls file /grant *S-1-1-0:(D,WDAC)
Will grant the user defined by sid S-1-1-0 Delete and Write DAC permissions to file.
```

# DSACLS

Displays or modifies permissions (ACLS) of an Active Directory Domain Services (AD DS) Object

```
        DSACLS object [/I:TSP] [/N] [/P:YN] [/G <group/user>:<perms> [...]]
                      [/R <group/user> [...]] [/D <group/user>:<perms> [...]]
                      [/S] [/T] [/A] [/resetDefaultDACL] [/resetDefaultSACL]
                      [/takeOwnership] [/user:<userName>] [/passwd:<passwd> | *]
                      [/simple]

object              Path to the AD DS object for which to display or manipulate the ACLs

Path is the RFC 1779 format of the name, as in
        CN=John Doe,OU=Software,OU=Engineering,DC=Widget,DC=com
A specific AD DS can be denoted by prepending \\server[:port]\ to the object, as in
        \\ADSERVER\CN=John Doe,OU=Software,OU=Engineering,DC=Widget,DC=US
no options          displays the security on the object.
/I                  Inheritance flags:
                        T: This object and sub objects
                        S: Sub objects only
                        P: Propagate inheritable permissions one level only.
/N                  Replaces the current access on the object, instead of editing it.
/P                  Mark the object as protected
                        Y:Yes
                        N:No
                    If /P option is not present, current protection flag is maintained.
/G   <group/user>:<perms>
```

```
                            Grant specified group (or user) specified permissions.
                            See below for format of <group/user> and <perms>
/D  <group/user>:<perms>
                            Deny specified group (or user) specified permissions.
                            See below for format of <group/user> and <perms>
/R  <group/user>    Remove all permissions for the specified group (or user).
                            See below for format of <group/user>
/S                  Restore the security on the object to the default for that object class as defined in AD
                    DS Schema. This option works when dsacls is bound to NTDS. To restore default ACL of an
                    object in AD LDS use /resetDefaultDACL and /resetDefaultSACL options.
/T                  Restore the security on the tree of objects to the default for the object class.
                    This switch is valid only with the /S option.
/A                  When displaying the security on an AD DS object, display the auditing information as
                    well as the permissions and ownership information.
/resetDefaultDACL   Restore the DACL on the object to the default for that object class as defined in AD DS
                    Schema.
/resetDefaultSACL   Restore the SACL on the object to the default for that object class as defined in AD DS
                    Schema.
/takeOwnership      Take ownership of the object.
/domain:<domainName> Connect to ldap server using this domain account of the user.
/user:<userName>    Connect to ldap server using this user name. If this option is not used dsacls will bind
                    as the currently logged on user, using SSPI.
/passwd:<passwd> | * Passwd for the user account.
/simple             Bind to server using ldap simple bind. Note that the clear text password will be sent
                    over the wire.

<user/group>            should be in the following forms:
                            group@domain or domain\group
                            user@domain or domain\user
                            FQDN of the user or groupv
                            A string SID

<perms> should be in the following form:

    [Permission bits];[Object/Property];[Inherited Object Type]
  Permission bits can have the following values concatenated together:

    Generic Permissions
        GR      Generic Read
        GE      Generic Execute
        GW      Generic Write
        GA      Generic All

    Specific Permissions
        SD      Delete
        DT      Delete an object and all of it's children
        RC      Read security information
        WD      Change security information
        WO      Change owner information
        LC      List the children of an object
        CC      Create child object
        DC      Delete a child object
                For these two permissions, if [Object/Property] is not specified to define a specific
                child object type, they apply all types of child objects otherwise they apply to that
                specific child object type.
        WS      Write To Self (also known as Validated Write). There are 3 kinds of validated writes:
                Self-Membership (bf9679c0-0de6-11d0-a285-00aa003049e2)applied to Group object. It allows
                updating membershipof a group in terms of adding/removing to its own account.
                Example: (WS; bf9679c0-0de6-11d0-a285-00aa003049e2; AU)applied to group X, allows an
                Authenticated User to add/remove oneself to/from group X, but not anybody else.
                Validated-DNS-Host-Name (72e39547-7b18-11d1-adef-00c04fd8d5cd)applied to computer ob-
                ject. It allows updating the DNS host name attribute that is compliant with the
                computer name & domain name.
                Validated-SPN (f3a64788-5306-11d1-a9c5-0000f80367c1) applied to computer object: It al-
                lows updating the SPN attribute that is compliant to the DNS host name of the
                computer.
        WP      Write property.
        RP      Read property.
                For these two permissions, if [Object/Property] is not specified to define a specific
                property, they apply to all properties of the object otherwise they apply to that
                specific property of the object.
        CA      Control access right
                For this permission, if [Object/Property] is not specified to define the specific "ex-
                tended right" for control access, it applies to all control accesses meaningful on the
                object, otherwise it applies to the specific extended right for that object.
        LO      List the object access.  Can be used to grant list access to a specific object if
                List Children (LC) is not granted to the parent as well can denied on specific objects
                to hide those objects if the user/group has LC on the parent.
                NOTE:  AD DS does NOT enforce this permission by default, it has to be configured to
                start checking for this permission.
```

[Object/Property] must be the display name of the object type or the property. For example "user" is the display name for user objects and "telephone number" is the display name for telephone number property.

[Inherited Object Type] must be the display name of the object type that the permissions are expected to be inherited to. The permissions MUST be Inherit Only.

NOTE: This must only be used when defining object specific permissions that override the default permissions defined in the AD DS schema for that object type. USE THIS WITH CAUTION and ONLY IF YOU UNDERSTAND object specific permissions.

Examples of a valid <perms> would be:

SDRCWDWO;;user
means:
Delete, Read security information, Change security information and Change ownership permissions on objects of type "user".

CCDC;group;
means:
Create child and Delete child permissions to create/delete objects of type group.

RPWP;telephonenumber;
means:
read property and write property permissions on telephone number property

You can specify more than one user in a command.
The command completed successfully

## DSGET

This tool's commands display the selected properties of a specific object in the directory. The dsget commands:

    **dsget computer** - displays properties of computers in the directory.
    **dsget contact** - displays properties of contacts in the directory.
    **dsget subnet** - displays properties of subnets in the directory.
    **dsget group** - displays properties of groups in the directory.
    **dsget ou** - displays properties of ou's in the directory.
    **dsget server** - displays properties of servers in the directory.
    **dsget site** - displays properties of sites in the directory.
    **dsget user** - displays properties of users in the directory.
    **dsget quota** - displays properties of quotas in the directory.
    **dsget partition** - displays properties of partitions in the directory.

To display an arbitrary set of attributes of any given object in the directory use the dsquery * command (see examples below).
For help on a specific command, type "dsget <ObjectType> /?" where <ObjectType> is one of the supported object types shown above.
For example, dsget ou /?.

**Remarks**:
The dsget commands help you to view the properties of a specific object in the directory: the input to dsget is an object and the output is a list of properties for that object. To find all objects that meet a given search criterion, use the dsquery commands (dsquery /?).
The dsget commands support piping of input to allow you to pipe results from the dsquery commands as input to the dsget commands and display detailed information on the objects found by the dsquery commands.
Commas that are not used as separators in distinguished names must be escaped with the backslash ("\") character (for example, "CN=Company\, Inc.,CN=Users,DC=microsoft,DC=com").
Backslashes used in distinguished names must be escaped with a backslash (for example, "CN=Sales\\Latin America,OU=Distribution Lists,DC=microsoft,DC=com").

**Examples**:
To find all users with names starting with "John" and display their office numbers:
    **dsquery user -name John* | dsget user -office**
To display the sAMAccountName, userPrincipalName and department attributes of the object whose DN is ou=Test,dc=microsoft,dc=com:
    **dsquery * ou=Test,dc=microsoft,dc=com -scope base -attr sAMAccountName userPrincipalName department**
To read all attributes of any object use the dsquery * command.
For example, to read all attributes of the object whose DN is ou=Test,dc=microsoft,dc=com:
    **dsquery * ou=Test,dc=microsoft,dc=com -scope base -attr ***

## DSQUERY

This tool's commands suite allow you to query the directory according to specified criteria. Each of the following dsquery commands finds objects of a specific object type, with the exception of dsquery *, which can query for any type of object:

    dsquery computer - finds computers in the directory.
    dsquery contact - finds contacts in the directory.
    dsquery subnet - finds subnets in the directory.
    dsquery group - finds groups in the directory.
    dsquery ou - finds organizational units in the directory.
    dsquery site - finds sites in the directory.

```
dsquery server    - finds AD DCs/LDS instances in the directory.
dsquery user      - finds users in the directory.
dsquery quota     - finds quota specifications in the directory.
dsquery partition - finds partitions in the directory.
dsquery *         - finds any object in the directory by using a generic LDAP query.
```

For help on a specific command, type "dsquery <ObjectType> /?" where <ObjectType> is one of the supported object types shown above.
For example, dsquery ou /?.

**Remarks:**
   The dsquery commands help you find objects in the directory that match a specified search criterion: the input to dsquery is a search criterion and the output is a list of objects matching the search. To get the properties of a specific object, use the dsget commands (dsget /?).
   The results from a dsquery command can be piped as input to one of the other directory service command-line tools, such as dsmod, dsget, dsrm or dsmove.
   Commas that are not used as separators in distinguished names must be escaped with the backslash ("\") character (for example, "CN=Company\, Inc.,CN=Users,DC=microsoft,DC=com").
   Backslashes used in distinguished names must be escaped with a backslash (for example, "CN=Sales\\ Latin America,OU=Distribution Lists,DC=microsoft,DC=com").

**Examples:**
   To find all computers that have been inactive for the last four weeks and remove them from the directory:
        **dsquery computer -inactive 4 | dsrm**
   To find all users in the organizational unit "ou=Marketing,dc=microsoft,dc=com" and add them to the Marketing Staff group:
        **dsquery user ou=Marketing,dc=microsoft,dc=com | dsmod group
        "cn=Marketing Staff,ou=Marketing,dc=microsoft,dc=com" -addmbr**
   To find all users with names starting with "John" and display his office number:
        **dsquery user -name John* | dsget user -office**
   To display an arbitrary set of attributes of any given object in the directory use the dsquery * command. For example, to display the sAMAccountName, userPrincipalName and department attributes of the object whose DN is ou=Test,dc=microsoft,dc=com:
        **dsquery * ou=Test,dc=microsoft,dc=com -scope base -attr sAMAccountName userPrincipalName department**
   To read all attributes of the object whose DN is ou=Test,dc=microsoft,dc=com:
        **dsquery * ou=Test,dc=microsoft,dc=com -scope base -attr ***

## *DSQUERY COMPUTER /?*
   Finds computers in the directory matching specified search criteria.

```
dsquery computer [{<StartNode> | forestroot | domainroot}]
                 [-o {dn | rdn | samid}] [-scope {subtree | onelevel | base}]
                 [-name <Name>] [-desc <Description>] [-samid <SAMName>]
                 [-inactive <NumWeeks>] [-stalepwd <NumDays>] [-disabled]
                 [{-s <Server> | -d <Domain>}] [-u <UserName>]
                 [-p {<Password> | *}] [-q] [-gc]
                 [-limit <NumObjects>] [{-uc | -uco | -uci}]
```

Parameters:

| Value | Description |
|---|---|
| {<StartNode> \| forestroot \| domainroot} | The node where the search will start: forest root, domain root, or a node whose DN is <StartNode>. Can be "forestroot", "domainroot" or an object DN. If "forestroot" is specified, the search is done via the global catalog. Default: domainroot. |
| -o {dn \| rdn \| samid} | Specifies the output format. Default: distinguished name (DN). |
| -scope {subtree \| onelevel \| base} | Specifies the scope of the search: subtree rooted at start node (subtree); immediate children of start node only (onelevel); the base object represented by start node (base). Note that subtree and domain scope are essentially the same for any start node unless the start node represents a domain root. If forestroot is specified as <StartNode>, subtree is the only valid scope. Default: subtree. |
| -name <Name> | Finds computers whose name matches the value given by <Name>, e.g., "jon*" or "*ith" or "j*th". |
| -desc <Description> | Finds computers whose description matches the value given by <Description>, e.g., "jon*" or "*ith" or "j*th". |
| -samid <SAMName> | Finds computers whose SAM account name matches the filter given by <SAMName>. |
| -inactive <NumWeeks> | Finds computers that have been inactive (stale) for at least <NumWeeks> number of weeks. |
| -stalepwd <NumDays> | Finds computers that have not changed their password for at least <NumDays> number of days. |
| -disabled | Finds computers with disabled accounts. |
| {-s <Server> \| -d <Domain>} | -s <Server> connects to the AD DC/LDS instance with name <Server>. -d <Domain> connects to an AD DC in domain <Domain>. Default: an AD DC in the logon domain. |

| | |
|---|---|
| -u <UserName> | Connect as <UserName>. Default: the logged in user. User name can be: user name, domain\user name, or user principal name (UPN). |
| -p <Password> | Password for the user <UserName>. If * then prompt for password. |
| -q | Quiet mode: suppress all output to standard output. |
| -gc | Search in the Active Directory Domain Services global catalog. |
| -limit <NumObjects> | Specifies the number of objects matching the given criteria to be returned, where <NumObjects> is the number of objects to be returned. If the value of <NumObjects> is 0, all matching objects are returned. If this parameter is not specified, by default the first 100 results are displayed. |
| {-uc \| -uco \| -uci} | -uc  Specifies that input from or output to pipe is formatted in Unicode. -uco Specifies that output to pipe or file is formatted in Unicode. -uci Specifies that input from pipe or file is formatted in Unicode. |

**Remarks:**

The dsquery commands help you find objects in the directory that match a specified search criterion: the input to dsquery is a search criteria and the output is a list of objects matching the search. To get the properties of a specific object, use the dsget commands (dsget /?).

If a value that you supply contains spaces, use quotation marks around the text (for example, "CN=John Smith,CN=Users,DC=microsoft,DC=com").

If you enter multiple values, the values must be separated by spaces for example, a list of distinguished names).

**Examples:**

To find all computers in the current domain whose name starts with "ms" and whose description starts with "desktop", and display their DNs:
    **dsquery computer domainroot -name ms* -desc desktop***

To find all computers in the organizational unit (OU) givenby ou=sales,dc=microsoft,dc=com and display their DNs:
    **dsquery computer ou=sales,dc=microsoft,dc=com**

## *DSQUERY CONTACT /?*

Finds contacts per given criteria.

    dsquery contact [{<StartNode> | forestroot | domainroot}][-o {dn | rdn}]
                    [-scope {subtree | onelevel | base}] [-name <Name>] [-desc <Description>]
                    [{-s <Server> | -d <Domain>}] [-u <UserName>][-p {<Password> | *}] [-q] [-gc]
                    [-limit <NumObjects>] [{-uc | -uco | -uci}]

**Parameters:**

| Value | Description |
|---|---|
| {<StartNode> \| forestroot \| domainroot} | The node where the search will start: forest root, domain root, or a node whose DN is <StartNode>. Can be "forestroot", "domainroot" or an object DN. If "forestroot" is specified, the search is done via the global catalog. Default: domainroot. |
| -o {dn \| rdn} | Specifies the output format. Default: distinguished name (DN). |
| -scope {subtree \| onelevel \| base} | Specifies the scope of the search: subtree rooted at start node (subtree); immediate children of start node only (oneleel); the base object represented by start node (base). Note that subtree and domain scope are essentially the same for any start node unless the start node represents a domain root. If forestroot is specified as <StartNode>, subtree is the only valid scope. Default: subtree. |
| -name <Name> | Finds all contacts whose name matches the filter given by <Name>, e.g., "jon*" or *ith" or "j*th". |
| -desc <Description> | Finds contacts with descriptions matching the value given by <Description>, e.g., "corp*" or *branch" or "j*th". |
| {-s <Server> \| -d <Domain>} | **-s <Server>** connects to the AD DC/LDS instance with name <Server>. **-d <Domain>** connects to an AD DC in domain <Domain>. Default: an AD DC in the logon domain. |
| -u <UserName> | Connect as <UserName>. Default: the logged in user. User name can be: user name, domain\user name, or user principal name (UPN). |
| -p <Password> | Password for the user <UserName>. If * then prompt for password. |
| -q | Quiet mode: suppress all output to standard output. |
| -gc | Search in the Active Directory Domain Services global catalog. |
| -limit <NumObjects> | Specifies the number of objects matching the given criteria to be returned, where <NumObjects> is the number of objects to be returned. If the value of <NumObjects> is 0, all matching objects are returned. If this parameter is not specified, by default the first 100 results are displayed. |
| {-uc \| -uco \| -uci} | -uc Specifies that input from or output to pipe is formatted in Unicode. -uco Specifies that output to pipe or file is formatted in Unicode. -uci Specifies that input from pipe or file is formatted in Unicode. |

**Remarks:**

The dsquery commands help you find objects in the directory that match a specified search criterion: the input to dsquery is a search criteriaand the output is a list of objects matching the search. To get the properties of a specific object, use the dsget commands (dsget /?).

If a value that you supply contains spaces, use quotation marks around the text (for example, "CN=John Smith,CN=Users,DC=microsoft,DC=com").
If you enter multiple values, the values must be separated by spaces (for example, a list of distinguished names).

## DSQUERY SUBNET /?
Finds subnets in the directory per given criteria.

```
dsquery subnet      [-o {dn | rdn}] [-name <Name>][-desc <Description>] [-loc <Location>]
                    [-site <SiteName>][{-s <Server> | -d <Domain>}] [-u <UserName>]
                    [-p {<Password> | *}] [-q] [-gc] [-limit <NumObjects>] [{-uc | -uco | -uci}]
```

### Parameters:

| Value | Description |
|---|---|
| -o {dn | rdn} | Specifies the output format.<br>Default: distinguished name (DN). |
| -name <Name> | Find subnets whose name matches the value given by <Name>, e.g., "10.23.*" or "12.2.*". |
| -desc <Description> | Find subnets whose description matches the value given by <Description>, e.g., "corp*" or "*nch" or "j*th". |
| -loc <Location> | Find subnets whose location matches the value given by <Location>. |
| -site <SiteName> | Find subnets that are part of site <SiteName>. |
| {-s <Server> | -d <Domain>} | -s <Server> connects to the AD DC/LDS instance with name <Server>.<br>-d <Domain> connects to an AD DC in domain <Domain>.<br>Default: an AD DC in the logon domain. |
| -u <UserName> | Connect as <UserName>. Default: the logged in user. User name can be: user name, domain\user name, or user principal name (UPN). |
| -p <Password> | Password for the user <UserName>. If * then prompt for password. |
| -q | Quiet mode: suppress all output to standard output. |
| -gc | Search in the Active Directory Domain Services global catalog. |
| -limit <NumObjects> | Specifies the number of objects matching the given criteria to be returned, where <NumObjects> is the number of objects to be returned.<br>If the value of <NumObjects> is 0, all matching objects are returned.<br>If this parameter is not specified, by default the first 100 results are displayed. |
| {-uc | -uco | -uci} | -uc Specifies that input from or output to pipe is formatted in Unicode.<br>-uco Specifies that output to pipe or file is formatted in Unicode.<br>-uci Specifies that input from pipe or file is formatted in Unicode. |

### Remarks:
The dsquery commands help you find objects in the directory that match a specified search criterion: the input to dsquery is a search criteria and the output is a list of objects matching the search. To get the properties of a specific object, use the dsget commands (dsget /?).
If a value that you supply contains spaces, use quotation marks around the text (for example, "CN=John Smith,CN=Users,DC=microsoft,DC=com").
If you enter multiple values, the values must be separated by spaces (for example, a list of distinguished names).

### Examples:
To find all subnets with the network IP address starting with 123.12:
    **dsquery subnet -name 123.12.***
To find all subnets in the site whose name is "Latin-America", and display their names as Relative Distinguished Names (RDNs):
    **dsquery subnet -o rdn -site Latin-America**
To list the names (RDNs) of all subnets defined in the directory:
    **dsquery subnet -o rdn**

## DSQUERY GROUP /?
Finds groups in the directory per given criteria.

```
dsquery group [{<StartNode> | forestroot | domainroot}][-o {dn | rdn | samid}]
              [-scope {subtree | onelevel | base}] [-name <Name>] [-desc <Description>]
              [-samid <SAMName>] [{-s <Server> | -d <Domain>}] [-u <UserName>]
              [-p {<Password> | *}] [-q] [-gc] [-limit <NumObjects>] [{-uc | -uco | -uci}]
```

### Parameters:

| Value | Description |
|---|---|
| {<StartNode> | forestroot | domainroot} | The node where the search will start: forest root, domain root, or a node whose DN is <StartNode>.<br>Can be "forestroot", "domainroot" or an object DN. If "forestroot" is specified, the search is done via the global catalog.<br>Default: domainroot. |
| -o {dn | rdn | samid} | Specifies the output format.<br>Default: distinguished name (DN). |
| -scope {subtree | onelevel | base} | Specifies the scope of the search: subtree rooted at start node (subtree); immediate children of start node only (onelevel); the base object represented by start node (base).<br>Note that subtree and domain scope are essentially the same for any start node unless the start node represents a domain root.<br>If forestroot is specified as <StartNode>, subtree is the only valid scope. |

| | |
|---|---|
| -name <Name> | Default: subtree.<br>Find groups whose name matches the value given by <Name>, e.g., "jon*" or "*ith" or "j*th". |
| -desc <Description> | Find groups whose description matches the value given by <Description>, e.g., "jon*" or "*ith" or "j*th". |
| -samid <SAMName> | Find groups whose SAM account name matches the value given by <SAMName>. |
| {-s <Server> \| -d <Domain>} | -s <Server> connects to the AD DC/LDS instance with name <Server>.<br>-d <Domain> connects to an AD DC in domain <Domain>.<br>Default: an AD DC in the logon domain. |
| -u <UserName> | Connect as <UserName>. Default: the logged in user. User name can be: user name, domain\user name, or user principal name (UPN). |
| -p <Password> | Password for the user <UserName>.<br>If * is specified, then you are prompted for a password. |
| -q | Quiet mode: suppress all output to standard output. |
| -gc | Search in the Active Directory Domain Services global catalog. |
| -limit <NumObjects> | Specifies the number of objects matching the given criteria to be returned, where <NumObjects> is the number of objects to be returned.<br>If the value of <NumObjects> is 0, all matching objects are returned.<br>If this parameter is not specified, by default the first 100 results are displayed. |
| {-uc \| -uco \| -uci} | -uc Specifies that input from or output to pipe is formatted in Unicode.<br>-uco Specifies that output to pipe or file is formatted in Unicode.<br>-uci Specifies that input from pipe or file is formatted in Unicode. |

**Remarks:**

The dsquery commands help you find objects in the directory that match a specified search criterion: the input to dsquery is a search criteria and the output is a list of objects matching the search. To get the properties of a specific object, use the dsget commands (dsget /?).

If a value that you supply contains spaces, use quotation marks around the text (for example, "CN=John Smith,CN=Users,DC=microsoft,DC=com").

If you enter multiple values, the values must be separated by spaces(for example, a list of distinguished names).

**Examples:**

To find all groups in the current domain whose name starts with "ms" and whose description starts with "admin", and display their DNs:
    **dsquery group domainroot -name ms* -desc admin***
Find all groups in the domain given by dc=microsoft,dc=comand display their DNs:
    **dsquery group dc=microsoft,dc=com**

## DSQUERY OU /?

Finds organizational units (OUs) in the directory according to specified criteria.

    dsquery ou      [{<StartNode> | forestroot | domainroot}][-o {dn | rdn}]
                    [-scope {subtree | onelevel | base}] [-name <Name>] [-desc <Description>]
                    [{-s <Server> | -d <Domain>}] [-u <UserName>] [-p {<Password> | *}] [-q] [-gc]
                    [-limit <NumObjects>] [{-uc | -uco | -uci}]

**Parameters:**

| Value | Description |
|---|---|
| {<StartNode> \| forestroot \| domainroot} | The node where the search will start: forest root, domain root, or a node whose DN is <StartNode>.<br>Can be "forestroot", "domainroot" or an object DN.<br>If "forestroot" is specified, the search is done via the global catalog. Default: domainroot. |
| -o {dn \| rdn} | Specifies the output format.<br>Default: distinguished name (DN). |
| -scope {subtree \| onelevel \| base} | Specifies the scope of the search: subtree rooted at start node (subtree); immediate children of start node only (onelevel); the base object represented by start node (base).<br>Note that subtree and domain scope are essentially the same for any start node unless the start node represents a domain root.<br>If forestroot is specified as <StartNode>, subtree is the only valid scope.<br>Default: subtree. |
| -name <Name> | Find organizational units (OUs) whose name matches the value given by <Name>, e.g., "jon*" or "*ith" or "j*th". |
| -desc <Description> | Find OUs whose description matches the value given by <Description>, e.g., "jon*" or "*ith" or "j*th". |
| {-s <Server> \| -d <Domain>} | -s <Server> connects to the AD DC/LDS instance with name <Server>.<br>-d <Domain> connects to an AD DC in domain <Domain>.<br>Default: an AD DC in the logon domain. |
| -u <UserName> | Connect as <UserName>. Default: the logged in user. User name can be: user name, domain\user name, or user principal name (UPN). |
| -p <Password> | Password for the user <UserName>.<br>If * then prompt for password. |
| -q | Quiet mode: suppress all output to standard output. |
| -gc | Search in the Active Directory Domain Services global catalog. |

```
-limit <NumObjects>      Specifies the number of objects matching the given criteria to be returned, where
                         <NumObjects> is the number of objects to be returned.
                         If the value of <NumObjects> is 0, all matching objects are returned.
                         If this parameter is not specified, by default the first 100 results are displayed.
{-uc | -uco | -uci}      -uc  Specifies that input from or output to pipe is formatted in Unicode.
                         -uco Specifies that output to pipe or file is formatted in Unicode.
                         -uci Specifies that input from pipe or file is formatted in Unicode.
```

**Remarks:**
   The dsquery commands help you find objects in the directory that match a specified search criterion: the input to dsquery is a search criteria and the output is a list of objects matching the search. To get the properties of a specific object, use the dsget commands (dsget /?).
   If a value that you supply contains spaces, use quotation marks around the text (for example, "CN=John Smith,CN=Users,DC=microsoft,DC=com").
   If you enter multiple values, the values must be separated by spaces (for example, a list of distinguished names).

**Examples:**
   To find all OUs in the current domain whose name starts with "ms" and whose description starts with "sales", and display their DNs:
      **dsquery ou domainroot -name ms* -desc sales***
   To find all OUs in the domain given by dc=microsoft,dc=com and display their DNs:

      **dsquery ou dc=microsoft,dc=com**

## DSQUERY SITE
   Finds sites in the directory per given criteria.

```
   dsquery site [-o {dn | rdn}] [-name <Name>][-desc <Description>] [{-s <Server> | -d <Domain>}]
           [-u <UserName>] [-p {<Password> | *}] [-q][-gc] [-limit <NumObjects>] [{-uc | -uco | -uci}]
```

**Parameters:**
```
Value                    Description
-o {dn | rdn}            Specifies the output format.
                         Default: distinguished name (DN).
-name <Name>             Finds subnets whose name matches the value given by <Name>, e.g., "NA*" or "Europe*".
-desc <Description>      Finds subnets whose description matches the filter given by <Description>, e.g.,
                         "corp*" or "*nch" or "j*th".
{-s <Server> | -d <Domain>}
                         -s <Server> connects to the AD DC/LDS instance with name <Server>.
                         -d <Domain> connects to an AD DC in domain <Domain>.
                         Default: an AD DC in the logon domain.
-u <UserName>            Connect as <UserName>. Default: the logged in user. User name can be: user name, do-
                         main\user name, or user principal name (UPN).
-p <Password>            Password for the user <UserName>. If * then prompt for password.
-q                       Quiet mode: suppress all output to standard output.
-gc                      Search in the Active Directory Domain Services global catalog.
-limit <NumObjects>      Specifies the number of objects matching the given criteria to be returned, where
                         <NumObjects> is the number of objects to be returned.
                         If the value of <NumObjects> is 0, all matching objects are returned.
                         If this parameter is not specified, by default the first 100 results are diplayed.
{-uc | -uco | -uci}      -uc  Specifies that input from or output to pipe is formatted in Unicode.
                         -uco Specifies that output to pipe or file is formatted in Unicode.
                         -uci Specifies that input from pipe or file is formatted in Unicode.
```

**Remarks:**
   The dsquery commands help you find objects in the directory that match a specified search criterion: the input to dsquery is a search criteria and the output is a list of objects matching the search. To get the properties of a specific object, use the dsget commands (dsget /?).
   If a value that you supply contains spaces, use quotation marks around the text (for example, "CN=John Smith,CN=Users,DC=microsoft,DC=com").
   If you enter multiple values, the values must be separated by spaces (for example, a list of distinguished names).

**Examples:**
   To find all sites in North America with name starting with "north" and display their DNs:
      **dsquery site -name north***
   To list the distinguished names (RDNs) of all sites defined in the directory:
      **dsquery site -o rdn**

## DSQUERY SERVER /?
   Finds Active Directory Domain Controllers / Active Directory Lightweight Directory Services instances according to specified search criteria.

```
      dsquery server [-o {dn | rdn}] [-forest][-domain <DomainName>] [-site <SiteName>]
                 [-name <Name>] [-desc <Description>][-hasfsmo {schema | name | infr | pdc | rid}]
                 [-isgc] [-isreadonly] [{-s <Server> | -d <Domain>}] [-u <UserName>]
                 [-p {<Password> | *}] [-q] [-gc] [-limit <NumObjects>] [{-uc | -uco | -uci}]
```

**Parameters:**

Value                    Description

| | |
|---|---|
| -o {dn \| rdn} | Specifies output format.<br>Default: distinguished name (DN). |
| -forest | Finds all Active Directory Domain Controllers in the current forest. |
| -domain <DomainName> | Finds all AD DCs in the domain with a DNS name matching <DomainName>. |
| -site <SiteName> | Finds all AD DCs that are part of site <SiteName>. |
| -name <Name> | Finds AD DCs with names matching the value given by <Name>, e.g., "NA*" or "Europe*" or "j*th". |
| -desc <Description> | Finds AD DCs with descriptions matching the value given by <Description>, e.g., "corp*" or "j*th". |
| -hasfsmo {schema \| name \| infr \| pdc \| rid} | Finds the AD DC/LDS instance that holds the specified Flexible Single-master Operation (FSMO) role.<br>(For the "infr," "pdc" and "rid" FSMO roles, if no domain is specified with the domain parameter, the current domain is used.) AD LDS instances can have the schema and name FSMO roles. |
| -isgc | Find all AD DCs that are also global catalog servers (GCs) in the scope specified (if the -forest, -domain or -site parameters are not specified, then find all GCs in the current domain is used). |
| -isreadonly | Find all AD Read-only DCs in the scope specified (if the -forest, -domain or -site parameters are not specified, then find all RODCs in the current domain is used). |
| {-s <Server> \| -d <Domain>} | **-s <Server>** connects to the AD DC/LDS instance with name <Server>.<br>**-d <Domain>** connects to an AD DC in domain <Domain>.<br>Default: an AD DC in the logon domain. |
| -u <UserName> | Connect as <UserName>. Default: the logged in user. User name can be: user name, **domain\user name, or user principal name (UPN.** |
| -p <Password> | Password for the user <UserName>.<br>If * then prompt for password. |
| -q | Quiet mode: suppress all output to standard output. |
| -gc | Search in the Active Directory Domain Services global catalog. |
| -limit <NumObjects> | Specifies the number of objects matching the given criteria to be returned, where <NumObjects> is the number of objects to be returned. If the value of <NumObjects> is 0, all matching objects are returned.<br>If this parameter is not specified, by default the first 100 results are displayed. |
| {-uc \| -uco \| -uci} | **-uc** Specifies that input from or output to pipe is formatted in Unicode.<br>**-uco** Specifies that output to pipe or file is formatted in Unicode.<br>**-uci** Specifies that input from pipe or file is formatted in Unicode. |

**Remarks:**

The dsquery commands help you find objects in the directory that match a specified search criterion: the input to dsquery is a search criteria and the output is a list of objects matching the search. To get the properties of a specific object, use the dsget commands (dsget /?).

If a value that you supply contains spaces, use quotation marks around the text (for example, "CN=John Smith,CN=Users,DC=microsoft,DC=com").

If you enter multiple values, the values must be separated by spaces (for example, a list of distinguished names).

**Examples:**

To find all AD DCs in the current domain:
    **dsquery server**

To find all AD DCs in the forest and display their Relative Distinguished Names:
    **dsquery server -o rdn -forest**

To find all AD DCs in the site whose name is "Latin-America", and display their Relative Distinguished Names:
    **dsquery server -o rdn -site Latin-America**

Find the AD DC in the forest that holds the schema FSMO role:
    **dsquery server -forest -hasfsmo schema**

Find all AD DCs in the domain example.microsoft.com that are global catalog servers:
    **dsquery server -domain example.microsoft.com -isgc**

Find all AD DCs in the current domain that hold a copy of a given directory partition called "ApplicationSales":
    **dsquery server -part "Application*"**

## *DSQUERY USER*

Finds users in the directory per given criteria.

    **dsquery user** [{<StartNode> \| forestroot \| domainroot}][-o {dn \| rdn \| upn \| samid}]
            [-scope {subtree \| onelevel \| base}][-name <Name>] [-namep <Phonetic Name>]
            [-desc <Description>] [-upn <UPN>][-samid <SAMName>] [-inactive <NumWeeks>]
            [-stalepwd <NumDays>] [-disabled] [{-s <Server> \| -d <Domain>}] [-u <UserName>]
            [-p {<Password> \| *}] [-q] [-gc] [-limit <NumObjects>][{-uc \| -uco \| -uci}]

**Parameters:**

| Value | Description |
|---|---|
| {<StartNode> \| forestroot \| domainroot} | The node where the search will start:<br>forest root, domain root, or a node whose DN is <StartNode>.<br>Can be "forestroot", "domainroot" or an object DN. If "forestroot" is specified, the search is done via the global catalog.<br>Default: domainroot. |
| -o {dn \| rdn \| upn \| samid} | |

|   |   |
|---|---|
| | Specifies the output format.<br>Default: distinguished name (DN). |
| -scope {subtree \| onelevel \| base} | |
| | Specifies the scope of the search:<br>subtree rooted at start node (subtree); immediate children of start node only (onelevel); the base object represented by start node (base).<br>Note that subtree and domain scope are essentially the same for any start node unless the start node represents a domain root.<br>If forestroot is specified as <StartNode>, subtree is the only valid scope.<br>Default: subtree. |
| -name <Name> | Finds users whose name matches the filter given by <Name>, e.g., "jon*" or "*ith" or "j*th". |
| -namep <Phonetic Name> | Finds users whose phonetic display names are given by <Phonetic Name>, e.g., "▬╜╜╖" or "▬╖" or "▬*╖" |
| -desc <Description> | Finds users whose description matches the filter given by <Description>, e.g., "jon*" or "*ith" or "j*th". |
| -upn <UPN> | Finds users whose UPN matches the filter given by <UPN>. |
| -samid <SAMName> | Finds users whose SAM account name matches the filter given by <SAMName>. |
| -inactive <NumWeeks> | Finds users that have been inactive (not logged on) for at least <NumWeeks> number of weeks. |
| -stalepwd <NumDays> | Finds users that have not changed their password for at least <NumDays> number of days. |
| -disabled | Finds users whose account is disabled. |
| {-s <Server> \| -d <Domain>} | -s <Server> connects to the AD DC/LDS instance with name <Server>.<br>-d <Domain> connects to an AD DC in domain <Domain>.<br>Default: an AD DC in the logon domain. |
| -u <UserName> | Connect as <UserName>. Default: the logged in user. User name can be: user name, domain\user name, or user principal name (UPN). |
| -p <Password> | Password for the user <UserName>.<br>If * is specified, then you are prompted for a password. |
| -q | Quiet mode: suppress all output to standard output. |
| -gc | Search in the Active Directory Domain Services global catalog. |
| -limit <NumObjects> | Specifies the number of objects matching the given criteria to be returned, where <NumObjects> is the number of objects to be returned.<br>If the value of <NumObjects> is 0, all matching objects are returned. If this parameter is not specified, by default the first 100 results are displayed. |
| {-uc \| -uco \| -uci} | -uc   Specifies that input from or output to pipe is formatted in Unicode.<br>-uco  Specifies that output to pipe or file is formatted in Unicode.<br>-uci  Specifies that input from pipe or file is formatted in Unicode. |

### Remarks:
   The dsquery commands help you find objects in the directory that match a specified search criterion: the input to dsquery is a search criteria and the output is a list of objects matching the search. To get the properties of a specific object, use the dsget commands (dsget /?).
   If a value that you supply contains spaces, use quotation marks around the text (for example, "CN=John Smith,CN=Users,DC=microsoft,DC=com").
   If you enter multiple values, the values must be separated by spaces (for example, a list of distinguished names).

### Examples:
   To find all users in a given organizational unit (OU) whose name starts with "jon" and whose account has been disabled for logon and display their user principal names (UPNs):
      **dsquery user ou=Test,dc=microsoft,dc=com -o upn -name jon* -disabled**
   To find all users in only the current domain, whose names end with "smith" and who have been inactive for 3 weeks or more, and display their DNs:
      **dsquery user domainroot -name *smith -inactive 3**
   To find all users in the OU given by ou=sales,dc=microsoft,dc=com and display their UPNs:
      **dsquery user ou=sales,dc=microsoft,dc=com -o upn**

## *DSQUERY QUOTA*
   Quota specifications in the directory that match the specified search criteria. A quota specification determines the maximum number of directory objects a given security principal can own in a specific directory partition. If the predefined search criteria in this command is insufficient, then use the more general version of the query command, dsquery *.

      dsquery quota   [{domainroot | <ObjectDN>}] [-o {dn | rdn}] [-acct <Name>] [-qlimit <Filter>]
                      [-desc <Description>] [{-s <Server> | -d <Domain>}] [-u <UserName>]
                      [-p {<Password> | *}][-q] [-limit <NumberOfObjects>] [{-uc | -uco | -uci}]

### Parameters:

| Value | Description |
|---|---|
| {domainroot \| <ObjectDN>} | |
| | Specifies where the search should begin.<br>Use ObjectDN to specify the distinguished name (also known as DN), or use domainroot to specify the root of the current domain. Default: domainRoot |
| -o {dn \| rdn} | Specifies the output format. The default format is distinguished name (dn). |
| -acct <Name> | Finds the quota specifications assigned to the security principal (user, group, computer, or InetOrgPerson) as represented by Name. The -acct option can be provided in the form of the distinguished name of the security principal or the Domain\SAMAccountName of the security principal. |
| -qlimit <Filter> | Finds the quota specifications whose limit matches Filter. |

```
-desc <Description>        Searches for quota specifications that have a description attribute that matches
                           Description (for example, "jon*" or "*ith" or "j*th").
{-s <Server> | -d <Domain>}
                           Connects to a specified AD DC/LDS instance or domain.
                           By default, the computer is connected to an AD DC in the logon domain.
-u <UserName>              Specifies the user name with which the user logs on to a remote server. By de-
                           fault, -u uses the user name with which the user logged on. You can use any of the
                           following formats to specify a user name:
                               user name (for example, Linda)
                               domain\user name (for example, widgets\Linda)
                               user principal name (UPN) (for example, Linda@widgets.microsoft.com)
-p {<Password> | *}        Specifies to use either a password or a * to log on to a remote server. If you
                           type *, you are prompted for a password.
-q                         Suppresses all output to standard output (quiet mode).
-limit <NumberOfObjects>
                           Specifies the number of objects that match the given criteria to be returned. If
                           the value of NumberOfObjects is 0, all matching objects are returned. If this pa-
                           rameter is not specified, the first 100 results are displayed by default.
{-uc | -uco | -uci}        Specifies that output or input data is formatted in Unicode, as follows:
                           -uc    Specifies a Unicode format for input from or output to a pipe (|).
                           -uco   Specifies a Unicode format for output to a pipe (|) or a file.
                           -uci   Specifies a Unicode format for input from a pipe (|) or a file.
```

**Remarks:**
The results from a dsquery search can be piped as input to one of the other directory service command-line tools, such as dsget, dsmod, dsmove, dsrm, or to an additional dsquery search.
If a value that you use contains spaces, use quotation marks around the text(for example, "CN=Linda,CN=Users,DC=Microsoft,DC=Com").
If you use multiple values for a parameter, use spaces to separate the values(for example, a list of distinguished names).
If you do not specify any search filter options (that is, -forest, -domain,-site, -name, -desc, -hasfsmo, -isgc), the default search criterion is to find all servers in the current domain, as represented by an appropriate LDAP search filter.
When you specify values for Description, you can use the wildcard character (*) (for example, "NA*," "*BR," and "NA*BA").
Any value for Filter that you specify with qlimit is read as a string.
You must always use quotation marks around this parameter. Any value ranges you specify using <=, =, or >= must also be inside quotation marks (for example, -qlimit "=100", -qlimit "<=99", -qlimit ">=101").
To find quotas with no limit, use "-1". To find all quotas not equal to unlimited, use ">=-1".

**Examples:**
To list all of the quota specifications in the current domain, type:
    **dsquery quota domainroot**
To list all users whose name begins with "Jon" that have quotas assigned to them, type:
    **dsquery user -name jon* | dsquery quota domainroot -acct |**
    **dsget quota -acct**

## DSQUERY PARTITION /?

Finds partition objects in the directory that match the specified search criteria. If the predefined search criteria in this command is insufficient, then use the more general version of the query command, dsquery *.

```
            dsquery partition  [-o {dn | rdn}] [-part <Filter>][-desc <Description>]
                               [{-s <Server> | -d <Domain>}] [-u <UserName>][-p {<Password> | *}]
                               [-q] [-limit <NumberOfObjects>][{-uc | -uco | -uci}]
```

**Parameters:**

```
Value                      Description
-o {dn | rdn}              Specifies the output format. The default format is distinguished name (dn).
-part <Filter>             Finds partition specifications whose common name (CN) matches the filter given by
                           Filter.
{-s <Server> | -d <Domain>}
                           Connects to a specified AD DC/LDS instance or domain.
                           By default, the computer is connected to an AD DC in the logon domain.
-u <UserName>              Specifies the user name with which the user logs on to a remote server. By default,
                           -u uses the user name with which the user logged on. You can use any of the follow-
                           ing formats to specify a user name:
                               user name (for example, Linda)domain\user name (for example, widgets\Linda)
                               user principal name (UPN)(for example, Linda@widgets.microsoft.com)
-p {<Password> | *}        Specifies to use either a password or a * to log on to a remote server. If you type
                           *, you are prompted for a password.
-q                         Suppresses all output to standard output (quiet mode).
-limit <NumberOfObjects>
                           Specifies the number of objects that match the given criteria to be returned. If the
                           value of NumberOfObjects is 0, all matching objects are returned. If this parameter
                           is not specified, the first 100 results are displayed by default.
{-uc | -uco | -uci}        Specifies that output or input data is formatted in Unicode, as follows:
                           -uc    Specifies a Unicode format for input from or output to a pipe (|).
                           -uco   Specifies a Unicode format for output to a pipe (|) or a file.
                           -uci   Specifies a Unicode format for input from a pipe (|) or a file.
```

**Remarks:**

The results from a dsquery search can be piped as input to one of the otherdirectory service command-line tools, such as dsget, dsmod, dsmove, dsrm, or to an additional dsquery search.

If a value that you use contains spaces, use quotation marks around the text(for example, "CN=Linda,CN=Users,DC=Microsoft,DC=Com").
If you use multiple values for a parameter, use spaces to separate the values (for example, a list of distinguished names).
If you do not specify any search filter options (that is, -forest, -domain, -site, -name, -desc, -hasfsmo, -isgc), the default search criterion is to find all servers in the current domain, as represented by an appropriate LDAP search filter.
When you specify values for Description, you can use the wildcard character(*) (for example, "NA*," "*BR," and "NA*BA").

**Examples:**
To list the DNs of all directory partitions in the forest, type:
**dsquery partition**
To list the DNs of all directory partitions in the forest whose common names start with SQL, type:
**dsquery partition -part SQL***

# FSUTIL

```
---- Commands Supported ----

8dot3name       8dot3name management
behavior        Control file system behavior
dirty           Manage volume dirty bit
file            File specific commands
fsinfo          File system information
hardlink        Hardlink management
objectid        Object ID management
quota           Quota management
repair          Self healing management
reparsepoint    Reparse point management
resource        Transactional Resource Manager management
sparse          Sparse file control
tiering         Storage tiering property management
transaction     Transaction management
usn             USN management
volume          Volume management
```

# FSUTIL FSINFO

```
---- FSINFO Commands Supported ----

drives          List all drives
driveType       Query drive type for a drive
ntfsInfo        Query NTFS specific volume information
sectorInfo      Query sector information
statistics      Query file system statistics
volumeInfo      Query volume information
```

# FSUTIL QUOTA

```
C:\Windows\system32>fsutil quota
---- QUOTA Commands Supported ----

disable         Disable quota tracking and enforcement
enforce         Enable quota enforcement
modify          Sets disk quota for a user
query           Query disk quotas
track           Enable quota tracking
violations      Display quota violations
```

# FSUTIL QUOTA MODIFY HELP

```
Usage : fsutil quota modify <volume pathname> <threshold> <limit> <user>
   Eg : fsutil quota modify c: 3000 5000 domain\user
```

# GETMAC

This tool enables an administrator to display the MAC address for network adapters on a system.

```
GETMAC [/S system [/U username [/P [password]]]] [/FO format] [/NH] [/V]
```

Parameter List:
```
/S      system              Specifies the remote system to connect to.
/U      [domain\]user       Specifies the user context under which the command should execute.
/P      [password]          Specifies the password for the given user context. Prompts for input if omitted.
/FO     format              Specifies the format in which the output is to be displayed.
                            Valid values: "TABLE", "LIST", "CSV".
/NH                         Specifies that the "Column Header" should not be displayed in the output.
                            Valid only for TABLE and CSV formats.
/V                          Specifies that verbose output is displayed.
/?                          Displays this help message.
```

```
Examples:
    GETMAC /?
    GETMAC /FO csv
    GETMAC /S system /NH /V
    GETMAC /S system /U user
    GETMAC /S system /U domain\user /P password /FO list /V
    GETMAC /S system /U domain\user /P password /FO table /NH
```

## NBTSTAT

Displays protocol statistics and current TCP/IP connections using NBT (NetBIOS over TCP/IP).

```
NBTSTAT [ [-a RemoteName] [-A IP address] [-c] [-n] [-r] [-R] [-RR] [-s] [-S] [interval] ]

  -a   (adapter status)  Lists the remote machine's name table given its name
  -A   (Adapter status)  Lists the remote machine's name table given its IP address.
  -c   (cache)           Lists NBT's cache of remote [machine] names and their IP addresses
  -n   (names)           Lists local NetBIOS names.
  -r   (resolved)        Lists names resolved by broadcast and via WINS
  -R   (Reload)          Purges and reloads the remote cache name table
  -S   (Sessions)        Lists sessions table with the destination IP addresses
  -s   (sessions)        Lists sessions table converting destination IP addresses to computer NETBIOS names.
  -RR  (ReleaseRefresh)  Sends Name Release packets to WINS and then, starts Refresh

  RemoteName   Remote host machine name.
  IP address   Dotted decimal representation of the IP address.
  interval     Redisplays selected statistics, pausing interval seconds between each display. Press Ctrl+C to
               stop redisplaying statistics.
```

## NET

The syntax of this command is:

```
NET HELP
command
      -or-
NET command /HELP

   Commands available are:

   NET ACCOUNTS          NET HELPMSG         NET STATISTICS
   NET COMPUTER          NET LOCALGROUP      NET STOP
   NET CONFIG            NET PAUSE           NET TIME
   NET CONTINUE          NET SESSION         NET USE
   NET FILE              NET SHARE           NET USER
   NET GROUP             NET START           NET VIEW
   NET HELP

   NET HELP NAMES explains different types of names in NET HELP syntax lines.
   NET HELP SERVICES lists some of the services you can start.
   NET HELP SYNTAX explains how to read NET HELP syntax lines.
   NET HELP command | MORE displays Help one screen at a time.
```

## *NET ACCOUNTS*

**NET ACCOUNTS** updates the user accounts database and modifies password and logon requirements for all accounts.

When used without options, NET ACCOUNTS displays the current settings for password, logon limitations, and domain information.

```
        NET ACCOUNTS    [/FORCELOGOFF:{minutes | NO}] [/MINPWLEN:length] [/MAXPWAGE:{days | UNLIMITED}]
                        [/MINPWAGE:days][/UNIQUEPW:number] [/DOMAIN]
```

Two conditions are required in order for options used with NET ACCOUNTS to take effect:

   - The password and logon requirements are only effective if user accounts have been set up (use User Manager or the NET USER command).

   - The NetLogon service must be running on all servers in the domain that verify logon. NetLogon is started automatically when Windows starts.

**/FORCELOGOFF:{minutes | NO}**   Sets the number of minutes a user has before being forced to log off when the account expires or valid logon hours expire.
                                  NO, the default, prevents forced logoff.
**/MINPWLEN:length**              Sets the minimum number of characters for a password. The range is 0-14 characters; the default is 6 characters.
**/MAXPWAGE:{days | UNLIMITED}**  Sets the maximum number of days that a password is valid. No limit is specified by using UNLIMITED. /MAXPWAGE can't be less than /MINPWAGE. The range is 1-999; the default is to leave the value unchanged.
**/MINPWAGE:days**                Sets the minimum number of days that must pass before a user can change a password.
                                  A value of 0 sets no minimum time. The range is 0-999; the default is 0 days. /MINPWAGE can't be more than /MAXPWAGE.

| | |
|---|---|
| **/UNIQUEPW:number** | Requires that a user's passwords be unique through the specified number of password changes. The maximum value is 24. |
| **/DOMAIN** | Performs the operation on a domain controller of the current domain. Otherwise, the operation is performed on the local computer. |

**NET HELP command | MORE displays Help one screen at a time.**

## *NET COMPUTER*

NET COMPUTER adds or deletes computers from a domain database. This command is available only on Windows NT Servers.

    **NET COMPUTER  \\computername {/ADD | /DEL}**

| | |
|---|---|
| **\\computername** | Specifies the computer to add or delete from the domain. |
| **/ADD** | Adds the specified computer to the domain. |
| **/DEL** | Removes the specified computer from the domain. |

## *NET CONFIG*

NET CONFIG displays configuration information of the Workstation or Server service. When used without the SERVER or WORKSTATION switch, it displays a list of configurable services. To get help with configuring a service, type NET HELP CONFIG service.

    **NET CONFIG [SERVER | WORKSTATION]**

| | |
|---|---|
| **SERVER** | Displays information about the configuration of the Server service. |
| **WORKSTATION** | Displays information about the configuration of the Workstation service. |

## *NET CONTINUE*

NET CONTINUE reactivates a Windows service that has been suspended by NET PAUSE.

    ***NET CONTINUE service***

| | |
|---|---|
| **service** | Is the paused service.<br>For example, one of the following:<br>NETLOGON<br>SCHEDULE<br>SERVER<br>WORKSTATION |

## *NET FILE*

NET FILE closes a shared file and removes file locks. When used without options, it lists the open files on a server. The listing includes the identification number assigned to an open file, the pathname of the file, the username, and the number of locks on the file.

    **NET FILE [id [/CLOSE]]**

This command works only on computers running the Server service.

| | |
|---|---|
| **id** | Is the identification number of the file. |
| **/CLOSE** | Closes an open file and removes file locks. Type this command from the server where the file is shared. |

## *NET GROUP*

NET GROUP adds, displays, or modifies global groups on servers. Used without parameters, it displays the groupnames on the server.

    **NET GROUP    [groupname [/COMMENT:"text"]] [/DOMAIN]**
                  **groupname {/ADD [/COMMENT:"text"] | /DELETE}  [/DOMAIN]**
                  **groupname username [...] {/ADD | /DELETE} [/DOMAIN]**

| | |
|---|---|
| **groupname** | Is the name of the group to add, expand, or delete.<br>Supply only a groupname to view a list of users in a group. |
| **/COMMENT:"text"** | Adds a comment for a new or existing group. Enclose the text in quotation marks. |
| **/DOMAIN** | Performs the operation on a domain controller of the current domain. Otherwise, the operation is performed on the local computer. |
| **username[ ...]** | Lists one or more usernames to add to or remove from a group. Separate multiple username entries with a space. |
| **/ADD** | Adds a group, or adds a username to a group. |
| **/DELETE** | Removes a group, or removes a username from a group. |

## *NET HELPMSG*

NET HELPMSG displays information about Windows network messages (such as error, warning, and alert messages). When you type NET HELPMSG and the numerical error (for example, "net helpmsg 2182"), Windows tells you about the message and suggests actions you can take to solve the problem.

    **NET HELPMSG message#**

**message#**  Is the numerical Windows error with which you need help.

## NET LOCALGROUP

NET LOCALGROUP modifies local groups on computers. When used without options, it displays the local groups on the computer.

```
NET LOCALGROUP [groupname [/COMMENT:"text"]] [/DOMAIN]
               groupname {/ADD [/COMMENT:"text"] | /DELETE}  [/DOMAIN]
               groupname name [...] {/ADD | /DELETE} [/DOMAIN]
```

| | |
|---|---|
| groupname | Is the name of the local group to add, expand, or delete. Supply only a groupname to view a list of users or global groups in a local group. |
| /COMMENT:"text" | Adds a comment for a new or existing group. Enclose the text in quotation marks. |
| /DOMAIN | Performs the operation on the domain controller of the current domain. Otherwise, the operation is performed on the local workstation. |
| name [ ...] | Lists one or more usernames or groupnames to add or to remove from a local group. Separate multiple entries with a space. Names may be users or global groups, but not other local groups. If a user is from another domain, preface the username with the domain name (for example, SALES\RALPHR). |
| /ADD | Adds a groupname or username to a local group. An account must be established for users or global groups added to a local group with this command. |
| /DELETE | Removes a groupname or username from a local group. |

## NET PAUSE

NET PAUSE suspends a Windows service or resource. Pausing a service puts it on hold.

```
NET PAUSE service
```

service          Is the service to be paused.
For example, one of the following:
NETLOGON
SCHEDULE
SERVER
WORKSTATION

## NET SESSION

NET SESSION lists or disconnects sessions between the computer and other computers on the network. When used without options, it displays information about all sessions with the computer of current focus.

```
NET SESSION     [\\computername] [/DELETE] [/LIST]
```

This command works only on servers.

| | |
|---|---|
| \\computername | Lists the session information for the named computer. |
| /DELETE | Ends the session between the local computer and computername, and closes all open files on the computer for the session. If computername is omitted, all sessions are ended. |
| /LIST | Displays information in a list rather than a table. |

## NET SHARE

NET SHARE makes a server's resources available to network users. When used without options, it lists information about all resources beingshared on the computer. For each resource, Windows reports the devicename(s) or pathname(s) and a descriptive comment associated with it.

```
NET SHARE sharename
          sharename=drive:path [/GRANT:user,[READ | CHANGE | FULL]]
                              [/USERS:number | /UNLIMITED]
                              [/REMARK:"text"]
                              [/CACHE:Manual | Documents| Programs | BranchCache | None]
          sharename [/USERS:number | /UNLIMITED]
                    [/REMARK:"text"]
                    [/CACHE:Manual | Documents | Programs | BranchCache | None]
          {sharename | devicename | drive:path} /DELETE
          sharename \\computername /DELETE
```

| | |
|---|---|
| sharename | Is the network name of the shared resource. Type NET SHARE with a sharename only to display information about that share. |
| drive:path | Specifies the absolute path of the directory to be shared. |
| /GRANT:user,perm | Creates the share with a security descriptor that gives the requested permissions to the specified user.  This option may be used more than once to give share permissions to multiple users. |
| /USERS:number | Sets the maximum number of users who can simultaneously access the shared resource. |
| /UNLIMITED | Specifies an unlimited number of users can simultaneously access the shared resource |
| /REMARK:"text" | Adds a descriptive comment about the resource. Enclose the text in quotation marks. |
| devicename | Is one or more printers (LPT1: through LPT9:) shared by sharename. |
| /DELETE | Stops sharing the resource. |
| /CACHE:Manual | Enables manual client caching of programs and documents from this share |
| /CACHE:Documents | Enables automatic caching of documents from this share |
| /CACHE:Programs | Enables automatic caching of documents and programs from this share |
| /CACHE:BranchCache | Manual caching of documents with BranchCache enabled from this share |
| /CACHE:None | Disables caching from this share |

## NET START
    NET START lists running services.

        **NET START [service]**

    **service**   May include one of the following services:
      BROWSER
      DHCP CLIENT
      EVENTLOG
      FILE REPLICATION
      NETLOGON
      PLUG AND PLAY
      REMOTE ACCESS CONNECTION MANAGER
      ROUTING AND REMOTE ACCESS
      RPCSS
      SCHEDULE
      SERVER
      SPOOLER
      TCP/IP NETBIOS HELPER
      UPS
      WORKSTATION

   When typed at the command prompt, service names of two words or more must be enclosed in quotation marks. For example, NET START "DHCP Client" starts the DHCP Client service.

   NET START can also start services not provided with Windows.

## NET STATISTICS
   NET STATISTICS displays the statistics log for the local Workstation or Server service. Used without parameters, NET STATISTICS displays the services for which statistics are available.

        **NET STATISTICS [WORKSTATION | SERVER]**

    **SERVER**        Displays the Server service statistics.
    **WORKSTATION**   Displays the Workstation service statistics.

## NET STOP
NET STOP stops Windows services. Stopping a service cancels any network connections the service is using. Also, some services are dependent on others. Stopping one service can stop others.

        **NET STOP service**

Some services cannot be stopped.

**service**   May be one of the following services:
  BROWSER
  DHCP CLIENT
  FILE REPLICATION
  NETLOGON
  REMOTE ACCESS CONNECTION MANAGER
  ROUTING AND REMOTE ACCESS
  SCHEDULE
  SERVER
  SPOOLER
  TCP/IP NETBIOS HELPER
  UPS
  WORKSTATION

NET STOP can also stop services not provided with Windows.

## NET TIME
   NET TIME synchronizes the computer's clock with that of another computer or domain, or displays the time for a computer or domain. When used without options on a Windows Server domain, it displays the current date and time at the computer designated as the time server for the domain.

        **NET TIME [\\computername | /DOMAIN[:domainname] | /RTSDOMAIN[:domainname]] [/SET]**

    **\\computername**          Is the name of the computer you want to check or synchronize with.
    **/DOMAIN[:domainname]**    Specifies to synchronize the time from the Primary Domain Controller of domainname.
    **/RTSDOMAIN[:domainname]** Specifies to synchronize with a Reliable Time Server from domainname.
    **/SET**                    Synchronizes the computer's time with the time on the specified computer or domain.

   The /QUERYSNTP and /SETSNTP options have been deprecated. Please use w32tm.exe to configure the Windows Time Service.

## NET USE
   NET USE connects a computer to a shared resource or disconnects a computer from a shared resource. When used without options, it lists the computer's connections.

        **NET USE [devicename | *] [\\computername\sharename[\volume] [password | *]]**

```
                [/USER:[domainname\]username]
                [/USER:[dotted domain name\]username]
                [/USER:[username@dotted domain name]
                [/SMARTCARD]
                [/SAVECRED]
                [[/DELETE] | [/PERSISTENT:{YES | NO}]]
   NET USE {devicename | *} [password | *] /HOME
   NET USE [/PERSISTENT:{YES | NO}]
```

| | |
|---|---|
| devicename | Assigns a name to connect to the resource or specifies the device to be disconnected. There are two kinds of devicenames: disk drives (D: through Z:) and printers (LPT1: through LPT3:). Type an asterisk instead of a specific devicename to assign the next available devicename. |
| \\computername | Is the name of the computer controlling the shared resource. If the computername contains blank characters, enclose the double backslash (\\) and the computername in quotation marks (" "). The computername may be from 1 to 15 characters long. |
| \sharename | Is the network name of the shared resource. |
| \volume | Specifies a NetWare volume on the server. You must have Client Services for Netware (Windows Workstations) or Gateway Service for Netware (Windows Server) installed and running to connect to NetWare servers. |
| password | Is the password needed to access the shared resource. |
| * | Produces a prompt for the password. The password is not displayed when you type it at the password prompt. |
| /USER | Specifies a different username with which the connection is made. |
| domainname | Specifies another domain. If domain is omitted, the current logged on domain is used. |
| username | Specifies the username with which to logon. |
| /SMARTCARD | Specifies that the connection is to use credentials on a smart card. |
| /SAVECRED | Specifies that the username and password are to be saved. This switch is ignored unless the command prompts for username and password. |
| /HOME | Connects a user to their home directory. |
| /DELETE | Cancels a network connection and removes the connection from the list of persistent connections. |
| /PERSISTENT | Controls the use of persistent network connections. The default is the setting used last. |
| YES | Saves connections as they are made, and restores them at next logon. |
| NO | Does not save the connection being made or subsequent connections; existing connections will be restored at next logon. Use the /DELETE switch to remove persistent connections. |

## NET USER

NET USER creates and modifies user accounts on computers. When used without switches, it lists the user accounts for the computer. The user account information is stored in the user accounts database.

```
       NET USER [username [password | *] [options]] [/DOMAIN]
                username {password | *} /ADD [options] [/DOMAIN]
                username [/DELETE] [/DOMAIN]
                username [/TIMES:{times | ALL}]
                username [/ACTIVE: {YES | NO}]
```

| | |
|---|---|
| username | Is the name of the user account to add, delete, modify, or view. The name of the user account can have as many as 20 characters. |
| password | Assigns or changes a password for the user's account. A password must satisfy the minimum length set with the /MINPWLEN option of the NET ACCOUNTS command. It can have as many as 14 characters. |
| * | Produces a prompt for the password. The password is not displayed when you type it at a password prompt. |
| /DOMAIN | Performs the operation on a domain controller of the current domain. |
| /ADD | Adds a user account to the user accounts database. |
| /DELETE | Removes a user account from the user accounts database. |
| Options | Are as follows: |

| Options | Description |
|---|---|
| /ACTIVE:{YES | NO} | Activates or deactivates the account. If the account is not active, the user cannot access the server. The default is YES. |
| /COMMENT:"text" | Provides a descriptive comment about the user's account. Enclose the text in quotation marks. |
| /COUNTRYCODE:nnn | Uses the operating system country/region code to implement the specified language files for a user's help and error messages. A value of 0 signifies the default country/region code. |
| /EXPIRES:{date | NEVER} | Causes the account to expire if date is set. NEVER sets no time limit on the account. An expiration date is in the form mm/dd/yy(yy). Months can be a number, spelled out, or abbreviated with three letters. Year can be two or four numbers. Use slashes(/) (no spaces) to separate parts of the date. |
| /FULLNAME:"name" | Is a user's full name (rather than a username). Enclose the name in quotation marks. |
| /HOMEDIR:pathname | Sets the path for the user's home directory. The path must exist. |
| /PASSWORDCHG:{YES | NO} | Specifies whether users can change their own password. The default is YES. |
| /PASSWORDREQ:{YES | NO} | Specifies whether a user account must have a password. The default is YES. |

| | |
|---|---|
| /LOGONPASSWORDCHG:{YES\|NO} | Specifies whether user should change their own password at the next logon.The default is NO. |
| /PROFILEPATH[:path] | Sets a path for the user's logon profile. |
| /SCRIPTPATH:pathname | Is the location of the user's logon script. |
| /TIMES:{times \| ALL} | Is the logon hours. TIMES is expressed as day[-day][,day[-day]],time[-time][,time[-time]], limited to 1-hour increments. |
| | Days can be spelled out or abbreviated. |
| | Hours can be 12- or 24-hour notation. For 12-hour notation, use am, pm, a.m., or p.m. ALL means a user can always log on, and a blank value means a user can never log on. Separate day and time entries with a comma, and separate multiple day and time entries with a semicolon. |
| /USERCOMMENT:"text" | Lets an administrator add or change the User Comment for the account. |
| /WORKSTATIONS:{computername[,...] \| *} | |
| | Lists as many as eight computers from which a user can log on to the network. If /WORKSTATIONS has no list or if the list is *, the user can log on from any computer. |

## *NET VIEW*

NET VIEW displays a list of resources being shared on a computer. When used without options, it displays a list of computers in the current domain or network.

**NET VIEW** [\\computername [/CACHE] | [/ALL] | /DOMAIN[:domainname]]

| | |
|---|---|
| \\computername | Is a computer whose shared resources you want to view. |
| /DOMAIN:domainname | Specifies the domain for which you want to view the available computers. If domainname is omitted, displays all domains in the local area network. |
| /CACHE | Displays the offline client caching settings for the resources on the specified computer |
| /ALL | Displays all the shares including the $ shares |

## NETDOM

The syntax of this command is:

NETDOM HELP command
    -or-
NETDOM command /help

    Commands available are:

    NETDOM ADD           NETDOM RESETPWD      NETDOM RESET
    NETDOM COMPUTERNAME    NETDOM QUERY         NETDOM TRUST
    NETDOM HELP           NETDOM REMOVE       NETDOM VERIFY
    NETDOM JOIN           NETDOM MOVENT4BDC
    NETDOM MOVE           NETDOM RENAMECOMPUTER

    NETDOM HELP SYNTAX explains how to read NET HELP syntax lines.
    NETDOM HELP command | MORE displays Help one screen at a time.

**Note** that verbose output can be specified by including /VERBOSE with any of the above netdom commands.

    The command completed successfully..

## *NETDOM QUERY*

        **NETDOM [ ADD | COMPUTERNAME | HELP | JOIN | MOVE | QUERY | REMOVE |
            MOVENT4BDC | RENAMECOMPUTER | RESET | TRUST | VERIFY | RESETPWD ]**

Try "NETDOM HELP" for more information.

C:\Windows\system32>NETDOM HELP QUERY
The syntax of this command is:

        **NETDOM QUERY** [/Domain:domain] [/Server:server][/UserD:user] [/PasswordD:[password | *]]
            [/Verify] [/RESEt] [/Direct] [/SecurePasswordPrompt]
            **WORKSTATION | SERVER | DC | OU | PDC | FSMO | TRUST**

NETDOM QUERY Queries the domain for information

| | |
|---|---|
| /Domain | Specifies the domain on which to query for the information. |
| /UserD | User account used to make the connection with the domain specified by the /Domain argument. |
| /PasswordD | Password of the user account specified by /UserD. A * means to prompt for the password. |
| /Server | Name of a specific domain controller that should be used to perform the query. |
| /Verify | For computers, verifies that the secure channel between the computer and the domain controller is operating properly. |
| | For trusts, verifies that the the trust between domains is operating properly. Only outbound trust will be verified. The user must have domain administrator credentials to get correct verification results. |
| /RESEt | Resets the secure channel between the computer and the domain controller; valid only for computer enumeration. |

| | |
|---|---|
| /Direct | Applies only for a TRUST query, lists only the direct trust links and omits the domains indirectly trusted through transitive links. Do not use with /Verify. |
| /SecurePasswordPrompt | Use secure credentials popup to specify credentials. This option should be used when smartcard credentials need to be specified. This option is only in effect when the password value is supplied as * |
| WORKSTATION | Query the domain for the list of workstations |
| SERVER | Query the domain for the list of servers |
| DC | Query the domain for the list of Domain Controllers |
| OU | Query the domain for the list of Organizational Units under which the specified user can create a machine object |
| PDC | Query the domain for the current Primary Domain Controller |
| FSMO | Query the domain for the current list of FSMO owners |
| TRUST | Query the domain for the list of its trusts |

The trust verify command checks only direct, outbound, Windows trusts. To verify an inbound trust, use the NETDOM TRUST command which allows you to specify credentials for the trusting domain.

## *NETDOM RENAMECOMPUTER*

NETDOM RENAMECOMPUTER renames a computer. If the computer is joined to a domain, then the computer object in the domain is also renamed. Certain services, such as the Certificate Authority, rely on a fixed machine name.

If any services of this type are running on the target computer, then a computer name change would have an adverse impact. This command should not be used to rename a domain controller.

```
NETDOM RENAMECOMPUTER machine /NewName:new-name
            [/UserD:user [/PasswordD:[password | *]]]
            [/UserO:user [/PasswordO:[password | *]]]
            [/Force]
            [/REBoot[:Time in seconds]]
            [/SecurePasswordPrompt]
```

| | |
|---|---|
| machine | is the name of the workstation or member server to be renamed |
| /NewName | Specifies the new name for the computer. Both the DNS host label and the NetBIOS name are changed to new-name. If new-name is longer than 15 characters, the NetBIOS name is derived from the first 15 characters |
| /UserD | User account used to make the connection with the domain. The domain can be specified as "/ud:domain\user". If domain is omitted, then the computer's domain is assumed. |
| /PasswordD | Password of the user account specified by /UserD. A * means to prompt for the password |
| /UserO | User account used to make the connection with the machine to be renamed. If omitted, then the currently logged on user's account is used. The user's domain can be specified as "/uo:domain\user". If domain is omitted, then a local computer account is assumed. |
| /PasswordO | Password of the user account specified by /UserO. A * means to prompt for the password |
| /Force | As noted above, this command can adversely affect some services running on the computer. The user will be prompted for confirmation unless the /FORCE switch is specified. |
| /REBoot | Specifies that the machine should be shutdown and automatically rebooted after the Rename has completed. The number of seconds before automatic shutdown can also be provided. Default is 30 seconds |
| /SecurePasswordPrompt | Use secure credentials popup to specify credentials. This option should be used when smartcard credentials need to be specified. This option is only in effect when the password value is supplied as * |

## *NETSTAT (servidor)*

Displays protocol statistics and current TCP/IP network connections.

```
NETSTAT [-a] [-b] [-e] [-f] [-n] [-o] [-p proto] [-r] [-s] [-x] [-t] [interval]
```

| | |
|---|---|
| -a | Displays all connections and listening ports. |
| -b | Displays the executable involved in creating each connection or listening port. In some cases well-known executables host multiple independent components, and in these cases the sequence of components involved in creating the connection or listening port is displayed. In this case the executable name is in [] at the bottom, on top is the component it called, and so forth until TCP/IP was reached. Note that this option can be time-consuming and will fail unless you have sufficient permissions. |
| -e | Displays Ethernet statistics. This may be combined with the -s option. |
| -f | Displays Fully Qualified Domain Names (FQDN) for foreign addresses. |
| -n | Displays addresses and port numbers in numerical form. |
| -o | Displays the owning process ID associated with each connection. |
| -p proto | Shows connections for the protocol specified by proto; proto may be any of: TCP, UDP, TCPv6, or UDPv6. If used with the -s option to display per-protocol statistics, proto may be any of: IP, IPv6, ICMP, ICMPv6, TCP, TCPv6, UDP, or UDPv6. |
| -r | Displays the routing table. |
| -s | Displays per-protocol statistics. By default, statistics are shown for IP, IPv6, ICMP, ICMPv6, TCP, TCPv6, UDP, and UDPv6; the -p option may be used to specify a subset of the default. |
| -t | Displays the current connection offload state. |
| -x | Displays NetworkDirect connections, listeners, and shared endpoints. |
| -y | Displays the TCP connection template for all connections. |

    **interval**       Cannot be combined with the other options.
                   Redisplays selected statistics, pausing interval seconds between each display.  Press CTRL+C to stop redisplaying statistics.  If omitted, netstat will print the current configuration information once.

# NSLOOKUP

Usage:

```
       nslookup [-opt ...]             # interactive mode using default server
       nslookup [-opt ...] - server    # interactive mode using 'server'
       nslookup [-opt ...] host        # just look up 'host' using default server
       nslookup [-opt ...] host server # just look up 'host' using 'server'

> help
Commands:   (identifiers are shown in uppercase, [] means optional)
NAME            - print info about the host/domain NAME using default server
NAME1 NAME2     - as above, but use NAME2 as server
help or ?       - print info on common commands
set OPTION      - set an option
    all                    - print options, current server and host
    [no]debug              - print debugging information
    [no]d2                 - print exhaustive debugging information
    [no]defname            - append domain name to each query
    [no]recurse            - ask for recursive answer to query
    [no]search             - use domain search list
    [no]vc                 - always use a virtual circuit
    domain=NAME            - set default domain name to NAME
    srchlist=N1[/N2/.../N6] - set domain to N1 and search list to N1,N2, etc.
    root=NAME              - set root server to NAME
    retry=X                - set number of retries to X
    timeout=X              - set initial time-out interval to X seconds
    type=X                 - set query type (ex. A,AAAA,A+AAAA,ANY,CNAME,MX,NS,PTR,SOA,SRV)
    querytype=X            - same as type
    class=X                - set query class (ex. IN (Internet), ANY)
    [no]msxfr              - use MS fast zone transfer
    ixfrver=X              - current version to use in IXFR transfer request
server NAME     - set default server to NAME, using current default server
lserver NAME    - set default server to NAME, using initial server
root            - set current default server to the root
ls [opt] DOMAIN [> FILE] - list addresses in DOMAIN (optional: output to FILE)
    -a          -  list canonical names and aliases
    -d          -  list all records
    -t TYPE     -  list records of the given RFC record type (ex. A,CNAME,MX,NS, PTR etc.)
view FILE       - sort an 'ls' output file and view it with pg
exit            - exit the program
```

# PATHPING

```
       pathping [-g host-list] [-h maximum_hops] [-i address] [-n] [-p period] [-q num_queries]
                [-w timeout] [-4] [-6] target_name
```

Options:

```
    -g host-list     Loose source route along host-list.
    -h maximum_hops  Maximum number of hops to search for target.
    -i address       Use the specified source address.
    -n               Do not resolve addresses to hostnames.
    -p period        Wait period milliseconds between pings.
    -q num_queries   Number of queries per hop.
    -w timeout       Wait timeout milliseconds for each reply.
    -4               Force using IPv4.
    -6               Force using IPv6.
```

# PING

```
    ping     [-t] [-a] [-n count] [-l size] [-f] [-i TTL] [-v TOS] [-r count] [-s count]
             [[-j host-list] | [-k host-list]] [-w timeout] [-R] [-S srcaddr] [-c compartment] [-p]
             [-4] [-6] target_name
```

Options:

```
    -t               Ping the specified host until stopped.
                     To see statistics and continue - type Control-Break; To stop - type Control-C.
    -a               Resolve addresses to hostnames.
    -n count         Number of echo requests to send.
    -l size          Send buffer size.
    -f               Set Don't Fragment flag in packet (IPv4-only).
    -i TTL           Time To Live.
    -v TOS           Type Of Service (IPv4-only. This setting has been deprecated and has no effect on the type
                     of service field in the IP Header).
    -r count         Record route for count hops (IPv4-only).
    -s count         Timestamp for count hops (IPv4-only).
    -j host-list     Loose source route along host-list (IPv4-only).
    -k host-list     Strict source route along host-list (IPv4-only).
```

```
-w timeout     Timeout in milliseconds to wait for each reply.
-R             Use routing header to test reverse route also (IPv6-only).
               Per RFC 5095 the use of this routing header has been deprecated. Some systems may drop
               echo requests if this header is used.
-S srcaddr     Source address to use.
-c compartment Routing compartment identifier.
-p             Ping a Hyper-V Network Virtualization provider address.
-4             Force using IPv4.
-6             Force using IPv6.
```

# SET

Displays, sets, or removes cmd.exe environment variables.

```
SET [variable=[string]]
```

**variable**   Specifies the environment-variable name.
**string**     Specifies a series of characters to assign to the variable.

**Type SET without parameters to display the current environment variables.**

If Command Extensions are enabled SET changes as follows:

SET command invoked with just a variable name, no equal sign or value will display the value of all variables whose prefix matches the name given to the SET command. For example:

```
SET P
```

**Would display all variables that begin with the letter 'P'**

**SET command will set the ERRORLEVEL to 1 if the variable name is not found in the current environment.**
**SET command will not allow an equal sign to be part of the name of a variable.**

**Two new switches have been added to the SET command:**

```
SET /A expression
SET /P variable=[promptString]
```

The /A switch specifies that the string to the right of the equal sign is a numerical expression that is evaluated. The expression evaluator is pretty simple and supports the following operations, in decreasing order of precedence:

```
()                         - grouping
! ~ -                      - unary operators
* / %                      - arithmetic operators
+ -                        - arithmetic operators
<< >>                      - logical shift
&                          - bitwise and
^                          - bitwise exclusive or
|                          - bitwise or
= *= /= %= += -=           - assignment
  &= ^= |= <<= >>=
,                          - expression separator
```

If you use any of the logical or modulus operators, you will need to enclose the expression string in quotes. Any non-numeric strings in the expression are treated as environment variable names whose values are converted to numbers before using them. If an environment variable name is specified but is not defined in the current environment, then a value of zero is used. This allows you to do arithmetic with environment variable values without having to type all those % signs to get their values. If SET /A is executed from the command line outside of a command script, then it displays the final value of the expression. The assignment operator requires an environment variable name to the left of the assignment operator. Numeric values are decimal numbers, unless prefixed by 0x for hexadecimal numbers, and 0 for octal numbers.
   So 0x12 is the same as 18 is the same as 022. Please note that the octal notation can be confusing: 08 and 09 are not valid numbers because 8 and 9 are not valid octal digits.
   The /P switch allows you to set the value of a variable to a line of input entered by the user. Displays the specified promptString before reading the line of input. The promptString can be empty.

Environment variable substitution has been enhanced as follows:

```
%PATH:str1=str2%
```

Would expand the PATH environment variable, substituting each occurrence of "str1" in the expanded result with "str2". "str2" can be the empty string to effectively delete all occurrences of "str1" from the expanded output. "str1" can begin with an asterisk, in which case it will match everything from the beginning of the expanded output to the first occurrence of the remaining portion of str1.

**May also specify substrings for an expansion.**

```
%PATH:~10,5%
```

Would expand the PATH environment variable, and then use only the 5 characters that begin at the 11th (offset 10) character of the expanded result. If the length is not specified, then it defaults to the remainder of the variable value. If either number (offset or length) is negative, then the number used is the length of the environment variable value added to the offset or length specified.

    %PATH:~-10%

would extract the last 10 characters of the PATH variable.

    %PATH:~0,-2%

Would extract all but the last 2 characters of the PATH variable.

Finally, support for delayed environment variable expansion has beenadded. This support is always disabled by default, but may be enabled/disabled via the /V command line switch to CMD.EXE. See CMD /?

Delayed environment variable expansion is useful for getting around the limitations of the current expansion which happens when a line of text is read, not when it is executed. The following example demonstrates the problem with immediate variable expansion:

    set VAR=before
    if "%VAR%" == "before" (
        set VAR=after
        if "%VAR%" == "after" @echo If you see this, it worked
    )

Would never display the message, since the %VAR% in BOTH IF statements is substituted when the first IF statement is read, since it logically includes the body of the IF, which is a compound statement. So the IF inside the compound statement is really comparing "before" with "after" which will never be equal. Similarly, the following example will not work as expected:

    set LIST=
    for %i in (*) do set LIST=%LIST% %i
    echo %LIST%

In that it will NOT build up a list of files in the current directory, but instead will just set the LIST variable to the last file found.
Again, this is because the %LIST% is expanded just once when the FOR statement is read, and at that time the LIST variable is empty.
So the actual FOR loop we are executing is:

    for %i in (*) do set LIST= %i

Which just keeps setting LIST to the last file found.
Delayed environment variable expansion allows you to use a different character (the exclamation mark) to expand environment variables at execution time. If delayed variable expansion is enabled, the above examples could be written as follows to work as intended:

    set VAR=before
    if "%VAR%" == "before" (
        set VAR=after
        if "!VAR!" == "after" @echo If you see this, it worked
    )

    set LIST=
    for %i in (*) do set LIST=!LIST! %i
    echo %LIST%

If Command Extensions are enabled, then there are several dynamic environment variables that can be expanded but which don't show up in the list of variables displayed by SET. These variable values are computed dynamically each time the value of the variable is expanded.
If the user explicitly defines a variable with one of these names, then that definition will override the dynamic one described below:

        %CD%                    expands to the current directory string.
        %DATE%                  expands to current date using same format as DATE command.
        %TIME%                  expands to current time using same format as TIME command.
        %RANDOM%                expands to a random decimal number between 0 and 32767.
        %ERRORLEVEL%            expands to the current ERRORLEVEL value
        %CMDEXTVERSION%         expands to the current Command Processor Extensions version number.
        %CMDCMDLINE%            expands to the original command line that invoked the Command Processor.
        %HIGHESTNUMANODENUMBER% expands to the highest NUMA node number on this machine.

## SETX
Creates or modifies environment variables in the user or system environment. Can set variables based on arguments, regkeys or file input.

SetX has three ways of working:

            SETX [/S system [/U [domain\]user [/P [password]]]] var value [/M]
            SETX [/S system [/U [domain\]user [/P [password]]]] var /K regpath [/M]
            SETX [/S system [/U [domain\]user [/P [password]]]]
                    /F file {var {/A x,y | /R x,y string}[/M] | /X} [/D delimiters]

```
Parameter List:
    /S        system              Specifies the remote system to connect to.
    /U        [domain\]user       Specifies the user context under which the command should execute.
    /P        [password]          Specifies the password for the given user context. Prompts for input if omitted.
    var                           Specifies the environment variable to set.
    value                         Specifies a value to be assigned to the environment variable.
    /K        regpath             Specifies that the variable is set based on information from a registry key.
                                  Path should be specified in the format of hive\key\...\value. For example,
                                  HKEY_LOCAL_MACHINE\System\CurrentControlSet\
                                  Control\TimeZoneInformation\StandardName.
    /F        file                Specifies the filename of the text file to use.
    /A        x,y                 Specifies absolute file coordinates(line X, item Y) as parameters to search
                                  within the file.
    /R        x,y string          Specifies relative file coordinates with respect to "string" as the search parame-
                                  ters.
    /M                            Specifies that the variable should be set in the system wide (HKEY_LOCAL_MACHINE)
                                  environment. The default is to set the variable under the HKEY_CURRENT_USER
                                  environment.
    /X                            Displays file contents with x,y coordinates.
    /D        delimiters          Specifies additional delimiters such as "," or "\". The built-in delimiters are
                                  space, tab, carriage return, and linefeed. Any ASCII character can be used as an
                                  additional delimiter. The maximum number of delimiters, including the built-in de-
                                  limiters, is 15.
    /?                            Displays this help message.

NOTE:  1) SETX writes variables to the master environment in the registry.
       2) On a local system, variables created or modified by this tool will be available in future command
          windows but not in the current CMD.exe command window.
       3) On a remote system, variables created or modified by this tool will be available at the next logon
          session.
       4) The valid Registry Key data types are REG_DWORD, REG_EXPAND_SZ,REG_SZ, REG_MULTI_SZ.
       5) Supported hives:  HKEY_LOCAL_MACHINE (HKLM), HKEY_CURRENT_USER (HKCU).
       6) Delimiters are case sensitive.
       7) REG_DWORD values are extracted from the registry in decimal format.

Examples:
    SETX MACHINE COMPAQ
    SETX MACHINE "COMPAQ COMPUTER" /M
    SETX MYPATH "%PATH%"
    SETX MYPATH ~PATH~
    SETX /S system /U user /P password  MACHINE COMPAQ
    SETX /S system /U user /P password MYPATH ^%PATH^%
    SETX TZONE /K HKEY_LOCAL_MACHINE\System\CurrentControlSet\Control\TimeZoneInformation\StandardName
    SETX BUILD /K "HKEY_LOCAL_MACHINE\Software\Microsoft\Windows
        NT\CurrentVersion\CurrentBuildNumber" /M
    SETX /S system /U user /P password TZONE /K HKEY_LOCAL_MACHINE\
        System\CurrentControlSet\Control\TimeZoneInformation\StandardName
    SETX /S system /U user /P password  BUILD /K
        "HKEY_LOCAL_MACHINE\Software\Microsoft\Windows NT\CurrentVersion\CurrentBuildNumber" /M
    SETX /F ipconfig.out /X
    SETX IPADDR /F ipconfig.out /A 5,11
    SETX OCTET1 /F ipconfig.out /A 5,3 /D "#$*."
    SETX IPGATEWAY /F ipconfig.out /R 0,7 Gateway
    SETX /S system /U user /P password  /F c:\ipconfig.out /X
```

## TRACERT

Usage: **tracert [-d] [-h maximum_hops] [-j host-list] [-w timeout] [-R] [-S srcaddr] [-4] [-6] target_name**

Options:
```
    -d                      Do not resolve addresses to hostnames.
    -h maximum_hops         Maximum number of hops to search for target.
    -j host-list            Loose source route along host-list (IPv4-only).
    -w timeout              Wait timeout milliseconds for each reply.
    -R                      Trace round-trip path (IPv6-only).
    -S srcaddr              Source address to use (IPv6-only).
    -4                      Force using IPv4.
    -6                      Force using IPv6.
```

## WHOAMI

This utility can be used to get user name and group information along with the respective security identifiers (SID), claims, privileges, logon identifier (logon ID) for the current user on the local system. I.e. who is the current logged on user?
If no switch is specified, tool displays the user name in NTLM format (domain\username).

WhoAmI has three ways of working:

```
        WHOAMI [/UPN | /FQDN | /LOGONID]
        WHOAMI { [/USER] [/GROUPS] [/CLAIMS] [/PRIV] } [/FO format] [/NH]
        WHOAMI /ALL [/FO format] [/NH]
```

Parameter List:
    /UPN                        Displays the user name in User Principal Name (UPN) format.

| | | |
|---|---|---|
| /FQDN | | Displays the user name in Fully Qualified Distinguished Name (FQDN) format. |
| /USER | | Displays information on the current user along with the security identifier (SID). |
| /GROUPS | | Displays group membership for current user, type of account, security identifiers (SID) and attributes. |
| /CLAIMS | | Displays claims for current user, including claim name, flags, type and values. |
| /PRIV | | Displays security privileges of the current user. |
| /LOGONID | | Displays the logon ID of the current user. |
| /ALL | | Displays the current user name, groups belonged to along with the security identifiers (SID), claims and privileges for the current user access token. |
| /FO | format | Specifies the output format to be displayed. Valid values are TABLE, LIST, CSV. Column headings are not displayed with CSV format. Default format is TABLE. |
| /NH | | Specifies that the column header should not be displayed in the output. This is valid only for TABLE and CSV formats. |

**Examples:**
```
WHOAMI
WHOAMI /UPN
WHOAMI /FQDN
WHOAMI /LOGONID
WHOAMI /USER
WHOAMI /USER /FO LIST
WHOAMI /USER /FO CSV
WHOAMI /GROUPS
WHOAMI /GROUPS /FO CSV /NH
WHOAMI /CLAIMS
WHOAMI /CLAIMS /FO LIST
WHOAMI /PRIV
WHOAMI /PRIV /FO TABLE
WHOAMI /USER /GROUPS
WHOAMI /USER /GROUPS /CLAIMS /PRIV
WHOAMI /ALL
WHOAMI /ALL /FO LIST
WHOAMI /ALL /FO CSV /NH
WHOAMI /?
```

# GLOSARIO

| CONCEPTO | DESCRIPCIÓN |
|---|---|
| CCNA | Cisco Certified Network Associate |
| CISCO | (Cisco Systems, Inc.) global company with headquarters in San Jose, California (USA). Designs and sells technology and network services such as: routers (routers), switches (switches), hubs, firewalls, IP telephony products, network management software like CiscoWorks, equipment Storage Area Networks. It was founded in 1984. |
| CSV | Comma-Separated Values, files stored tabular data in text format, separated by commas. To save your spreadsheet as a .csv file |
| DHCP | Dynamic Host Configuration Protocol. |
| DNS | Domain Name System. |
| EGP | Exterior Gateway Protocol. |
| EUI | Extended Unique Identifier (EUI-64). |
| FQDN | Fully Qualified Domain Name. |
| FSMO | Flexible Single Master Operation Roles. |
| HIPER-V | Program is a hypervisor-based virtualization for 64-bit systems based processors with AMD-V or Intel Virtualization Technology (management instruments can also be installed on x86). |
| ICMP/ICMPv6 | Internet Control Message Protocol. |
| IEEE | Institute of Electrical and Electronics Engineers. |
| IGP | Interior Gateway Protocol. |
| IP | Internet Protocol:IPv4, IPv6 |
| ISO | International Organization for Standardization. |
| LPTnúmero | Line Print Terminal. |
| NIC | Network Interface Card. |
| NTLM | NT LAN Manage. New Technology LAN Manage. |
| OSI | Open System Interconnection. |
| OSPF | Open Shortest Path First. |
| OUI | Organizationally Unique Identifier. |
| PID | Process IDentifier. |
| PROMPT | Indicative Point System, also called: Indicative system line, interactive point system command line (system). |
| RIP | Routing Information Protocol. |
| RPC | Remote Procedure Call. |
| SID | Security IDentifier, is a number used to identify user accounts, groups and Windows computers. |
| TCP/IP | Transmission Control Protocol/ Internet Protocol. |
| TCP/TCPv6 | Transmission Control Protocol (TCP) o Protocolo de Control de Transmisión. |
| UDP/UDPv6 | User Datagram Protocol, Protocolo de Datagramas de Usuario. Version 4 y version 6. |
| UPN | Universal Principal Name. |
| WEB | World Wide Web (WWW). |
| ICANN | Internet Corporation for Assigned Names and Numbers. |

# WEB REFERENCES

http://ss64.com/nt/
http://es.ccm.net/
https://technet.miscrosoft.com
https://es.wikipedia.org
http://www.gestion.org
http://www.pleplando.com/
http://www.pesadillo.com/
https://support.microsoft.com/es-es/kb/137984
http://www.xatakaon.com/

# REFERENCE COMMANDS

| Command | Page |
|---|---|
| CACLS | 48 |
| DSADD | 45,46 |
| DSGET | 51,52,53,54,100,101,106 |
| DSQUERY | 42,43 |
| DSQUERY COMPUTER | 42 |
| DSQUERY CONTACT | 42 |
| DSQUERY OU | 43 |
| DSQUERY PARTITION | 43,44 |
| DSQUERY QUOTA | 58,61,101,108 |
| DSQUERY SERVER | 43 |
| DSQUERY SITE | 43 |
| DSQUERY USER | 43,51 |
| FSUTIL | 61,62,63,110 |
| GETMAC | 10 |
| GPRESULT | 65,66,67 |
| GPUPDATE | 65 |
| HOSTNAME | 10 |
| ICACLS | 48,49,97,98 |
| IPCONFIG | 20,21,24,25,88 |
| NBTSTAT | 76,77,78,79,80,81,82,110 |
| NET ACCOUNTS | 65,113 |
| NET COMPUTER | 43,111 |
| NET CONFIG | 41,112 |
| NET CONTINUE | 41,42,112 |
| NET FILE | 40,112 |
| NET GROUP | 35,38 |
| NET HELPMSG | 37,40,65 |
| NET LOCCALGROUP | 35,36,112 |
| NET PAUSE | 42,113 |
| NET SESSION | 40,113 |
| NET SHARE | 32,34,39 |
| NET TIME | 42,114 |
| NET USE | 32,33 |
| NET USER | 37,38,115 |
| NET VIEW | 35,36,116 |
| NETDOM | 54,55,116 |
| NETSTAT | 69,71,72,73,75,76,117 |
| NSLOOKUP | 85,87,117 |
| PATHPING | 118 |
| PING | 14,15,16,17,18,19,20,85,118 |
| SET | 59,118,119,120 |
| SETX | 60,120 |
| TRACERT | 62,83,84,121 |
| WHOAMI | 27,28,29,42,57,121 |